Miguel de Benito Delgado

Effective two dimensional theories for multi-layered plates

λογος

Augsburger Schriften zur Mathematik, Physik und Informatik
Band 37

Edited by:
Professor Dr. B. Schmidt
Professor Dr. B. Aulbach
Professor Dr. F. Pukelsheim
Professor Dr. W. Reif
Professor Dr. D. Vollhardt

All the code is available at the authors online repositories.

Bibliographic information published by the Deutsche Nationalbibliothek

The Deutsche Nationalbibliothek lists this publication in the
Deutsche Nationalbibliografie; detailed bibliographic data are
available in the Internet at http://dnb.d-nb.de .

ISBN 978-3-8325-4984-8
ISSN 1611-4256

Logos Verlag Berlin GmbH
Comeniushof, Gubener Str. 47,
10243 Berlin
Tel.: +49 030 42 85 10 90
Fax: +49 030 42 85 10 92
INTERNET: http://www.logos-verlag.de

Effective two dimensional theories for multi-layered plates

Dissertation

zur Erlangung des akademischen Grades

Dr. rer. nat.

eingereicht an der

Mathematisch-Naturwissenschaftlich-Technischen Fakultät

der Universität Augsburg

von

Miguel de Benito Delgado

Augsburg, April 2019

Erstgutachter: Prof. Dr. Bernd Schmidt
Zweitgutachter: Prof. Dr. Malte A. Peter

Datum der mündlichen Prüfung: 2. Juli 2019

«All models are wrong, but some are useful.»

Box, G. E. P. (1979),
"Robustness in the strategy of scientific model building"

Contents

1

Lower dimensional models in elasticity

With the purpose of fixing notation and nomenclature, we begin by quickly reviewing some fundamental notions in elasticity theory.[1.1] We then discuss dimension reduction in this context and its mathematical justification. We continue with a brief review of the literature where Γ-convergence is applied for this purpose, to conclude with an outline of the present work and some acknowledgements. Please refer to Appendix B for the notation used throughout this work.

1.1 Elasticity, in a rush

The objects of study are a three dimensional **body** identified with an open, bounded and Lipschitz set $\Omega \subset \mathbb{R}^3$ and its **deformation** $y \colon \Omega \to \mathbb{R}^3$ under external forces or boundary conditions. When deformations can be assumed to be very small it is more convenient to use instead **displacements** $w \colon \Omega \to \mathbb{R}^3$, defined by $y(x) = x + w(x)$. Throughout we employ so-called **Lagrangian coordinates**, i.e. we track the deformations of material points wrt. the fixed domain Ω.[1.2]

Subject to external forces or boundary conditions, bodies deform. The fundamental assumption is that any deformation which is not a **rigid body motion** (the composition of a translation and a rotation) stores **elastic**

1.1. A thorough introduction to elasticity can be found in [Cia88], a gentle one from the perspective of differential geometry in [Cia05] and a deeper one in [MH94]. For a very good exposition of continuum mechanics with elasticity as an application see [TM05].

1.2. As opposed to the Eulerian description which instead tracks locations in space.

energy into the body which can be released after the extraneous conditions disappear and this release will bring the body back to its **reference configuration** Ω, without inducing any permanent alteration. If this does not hold, that is, in case the properties of the body are changed after the forces disappear, one can have **viscoelastic** or **plastic** behaviour, but we will not concern ourselves with these at all. If the reference configuration has zero elastic energy, we speak of a **natural state**. The elastic energy can be computed as the integral over Ω of a **stored energy density** W, which under mild assumptions turns out to be a function only of the position $x \in \Omega$ and the **deformation gradient** $\nabla y(x)$. When this is the case we speak of a **hyperelastic** material. The function W expresses the relationship between **strains** (local elongations and compressions in each direction) and **stresses** (internal forces induced by the strains). By our fundamental assumption above, W is non-negative and vanishes for rigid motions, or $W(x, \nabla y) = 0$ for all $\nabla y \in SO(3)$.

We model the strain by the change in metric induced by the map y in the body wrt. the flat metric, via the so-called **Green - St.Venant's tensor** $E(y) = \frac{1}{2}(\nabla^T y \, \nabla y - I)$. In terms of displacements $w = y - \mathrm{id}$, this is $E(w) = \frac{1}{2}(\nabla^T w + \nabla w + \nabla^T w \, \nabla w)$. Now we can characterise a **rigid motion** or **rigid body movement** as a deformation y such that $E(y) = 0$, i.e. $\nabla^T y \, \nabla y = I$, since there is no change in the distance between deformed points. The set of all rigid motions consists of all maps $x \mapsto Q\,x + c$ with $Q \in SO(3)$, $c \in \mathbb{R}^3$. Under the assumptions that displacements are "infinitesimally smaller" than the characteristic dimensions of the body, E is approximated by the **linear strain tensor** $e(w) := \nabla_s w = (\nabla^T w + \nabla w)/2$ and one speaks of **geometrically linear** elasticity.

Assuming a smooth energy density and a small displacement gradient $\|\nabla w\| \ll 1$, one can linearise the energy around the identity:

$$
\begin{aligned}
W(\nabla y) &= W(I) + DW(I)[\nabla w] + \tfrac{1}{2}D^2 W(I)[\nabla w, \nabla w] + h.o.t. \\
&\approx \tfrac{1}{2}D^2 W(I)[\nabla w, \nabla w] \\
&=: \tfrac{1}{2}Q_3(\nabla w),
\end{aligned}
$$

where we used that W vanishes on rigid motions so, in particular $W(I)$ and $DW(I)$ are zero, and where Q_3 is the **quadratic form of linear elas-**

ticity. In this setting we speak of **linearly elastic** materials. The form Q_3 vanishes exactly over the set of **linearised rigid motions**[1.3]

$$\mathscr{R} := \{x \mapsto Rx + b : R \in so(3), b \in \mathbb{R}^3\} = \{x \mapsto r \times x + b : r, b \in \mathbb{R}^3\},$$

where $so(3)$ is the space of antisymmetric matrices.

In order to define Q_3 in terms of the gradients ∇w one needs so-called **constitutive relations** between stresses and strains, which may take into account properties like **isotropy** (the body exhibits no "preferred direction" along which responses are different) and **homogeneity** (the body has the same behaviour at any material point $x \in \Omega$). The symmetries arising in isotropic, homogeneous materials imply that Q_3 has the form

$$Q_3(F) = \lambda \operatorname{tr}^2 F + 2\mu |F|^2$$

where $F = \nabla w \in \mathbb{R}^{3 \times 3}_{\text{sym}}$ is a strain tensor and λ, μ are the **Lamé constants** of the material.

There are several other couples of physically meaningful magnitudes related to these two constants, among which we mention **Young's modulus** E and **Poisson's ratio** v since we use them in the implementation of the discretisations. E is a measure of how the body extends or contracts in response to tensile or compressive stresses. v measures the tendency of materials to compress in directions perpendicular to the direction of elongation.[1.4]

1.3. In the setting of very small displacements, one must exclude symmetries (large displacements) from rigid motions, which means that the rotation matrices Q do not have the eigenvalue -1 and the maps $I + Q$ are invertible. Then we can define $R := (I - Q)(I + Q)^{-1}$ and recover Q with **Cayley's transform** $R \mapsto (I - R)(I + R)^{-1} = Q$. This bijection allows the identification of matrices Q with matrices R, so we can focus on maps $x \mapsto Rx + b$ with $R \in so(3)$. Additionally, each R is determined by just 3 coefficients, so there exists a vector $r \in \mathbb{R}^3$ such that $Rx + b = r \times x + b$.

1.4. E is defined as the quotients of stresses over strains along each direction, which reduces to a number for isotropic materials. Since strains are dimensionless, it has units of pressure N/m^2 or Pa, with typical values in the mega- and gigapascal range. v is the quotient of transverse strain to axial strain, with a sign, for each direction. Again, for isotropic materials this is only a number. Typical values range from 0 for materials with insignificant transversal expansion when compressed (e.g. cork) to 0.5 for incompressible ones (e.g. rubber), but materials have been designed beyond this range (*auxetic metamaterials*).

1.1.1 Some remarks on the energy density

Stored elastic energies require additional conditions to be physically relevant. An essential one is **frame invariance**, which expresses the fundamental idea that properties of physical processes should not depend on the observer. It is encoded as an invariance of the energy under maps in SO(3)

$$W(F) = W(RF) \text{ for all } R \in \text{SO}(3).$$

Note that frame invariance implies that W cannot be convex,[1.5] so that the problem of minimising the energy under, say, Dirichlet boundary conditions, may have no solution. This is however not an issue for the process of deriving limit theories using Γ-convergence because in the proofs it is only required that we have a "diagonally infimizing sequence", which is one very convenient feature of the method.

A second condition common in all of continuum mechanics is that of **non-interpenetration** of matter, encoded as the requirement that the energy be infinite whenever the *deformation gradient F* inverts the orientation of a region. In order to also avoid infinite compression of volumes it is actually required that[1.6]

$$W(F) = \infty \text{ if } \det F \leqslant 0 \text{ and } W(F) \to \infty \text{ as } \det F \to 0.$$

The simplest family of nonlinear hyperelastic models are the so-called **Green - St.Venant** materials. In these models the strain law is not linearised (geometrically non-linear), so that one uses Green - St.Venant's tensor, but the stress-strain relations are kept linear (linearly elastic). For the isotropic case in particular, this means a stored energy density

$$W(\nabla y) = \lambda \operatorname{tr}^2 E(y) + 2\mu |E(y)|^2.$$

This choice of W has the ugly property of violating non-interpenetration but also the desirable one of satisfying natural (from the technical point of view) *p*-growth conditions (for $p = 2$):

$$\begin{cases} W(F) \geqslant \alpha |F|^p - \beta, \\ W(F) \leqslant C(1 + |F|^p). \end{cases} \tag{1.1}$$

1.5. See e.g. [Cia88, Ex 3.7 and Thm 4.8-1].

1.6. One family of densities satisfying this condition while retaining other necessary technical properties (lower semicontinuity) consists of suitable **polyconvex** functions.

These provide (pre-)compactness of minimising sequences and are essential in many proofs of existence so they have been assumed throughout the literature. However they fail to be satisfied in other very important cases [Cia97, p. 349]. It is therefore of interest to relax conditions (1.1) in some way.

Our framework essentially requires the (inhomogeneous) energy to be frame invariant and bounded below by the distance to SO(3):[1.7]

$$W(x_3, F) \geqslant C \operatorname{dist}^2(F, \mathrm{SO}(3)),$$

plus some other technical conditions (Assumptions 2.2) related to the fact that W depends on the third spatial component. This places us in the non-convex setting, but with the potential to model physically relevant constraints like e.g. non-interpenetration.

1.2 Dimension reduction

Three-dimensional, non linearly elastic bodies under particular boundary conditions can be governed by complicated equations with no known analytic solutions. It is therefore fortunate that many physical applications exhibit a particularly simple structure, with one or two of the dimensions of the domain being relatively much larger than the other, or where

---

1.7. This lower bound also implies that $W_0(t, \cdot)$ cannot be convex: take for instance $A = \begin{pmatrix} 0 & -1 & 0 \\ 1 & 0 & 0 \\ 0 & 0 & 1 \end{pmatrix}$ and $B = \begin{pmatrix} 0 & 1 & 0 \\ -1 & 0 & 0 \\ 0 & 0 & 1 \end{pmatrix}$. Both are rotations but $\lambda A + (1 - \lambda) B \notin \mathrm{SO}(3)$. By the lower bound we have $W_0(t, \lambda A + (1 - \lambda) B) > 0 = \lambda W_0(t, A) + (1 - \lambda) W_0(t, B)$, that is: $W_0(t, \cdot)$ is not convex.

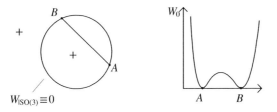

Fig. 1.1. Non convexity of $W_0(t, \cdot)$.

the internal characteristics of the bodies (isotropy, orthotropy, ...) or the loads they are subjected to (planar, compressive, ...) are such that kinematical and structural assumptions can be made which greatly simplify the problem without sacrificing too much accuracy. This reduced complexity can be translated to the equations, providing both ease of *interpretation* and *computation*: often, applications require not exact models, but *effective* ones, in the sense that they allow predicting how materials and structures behave under load, within acceptable error margins, with as little computation as possible and with a reasonable understanding of what failure modes can be and why. For centuries, analytically simple models have been employed for which analytic (approximate) solutions could be computed. Nowadays, even with vast computing resources available (by today's standards, anyway), many problems remain intractable if their dimension is not reduced. Examples of two dimensional models in elasticity are *plates* and *shells*, whereas *ribbons* and *rods* are typical one-dimensional models.

Given a true (physical, three dimensional) problem, the goal is to design a model, or **approximate problem** in a one- or two-dimensional setting, which within some parameter range and given accuracy approximates the original problem in the sense that any approximate solution retains the characteristics of the true one which are relevant to the application.

An important mathematical feature of dimension reduction is that questions like existence and regularity, or characterisations of minimal energy configurations are often possible in situations in which the three dimensional counterparts have proven to be elusive. Take for instance nonlinearly elastic, clamped (i.e. with Dirichlet boundary), St-Venant - Kirchhoff materials: because of the lack of convexity, minimisers are not known to exist in the general case, whereas in the 2D limit of e.g. elastic membranes, existence can be shown. A related difficulty is the derivation of sufficient conditions for minimisers to fulfill the Euler-Lagrange equations, again often easier in 2D than in 3D.

There are of course drawbacks to lower dimensional models. Obviously they only provide approximations to the real problems and there are generally no rigorous estimates of the error made, nor rigorous procedures to assess their validity in specific applications. Also numerical methods, which in principle greatly benefit from the reduced number of degrees of freedom, need to be adapted to avoid issues like *shear* and *membrane locking* or lack of convergence due to bogus boundary conditions or singularities in the solutions. We briefly touch upon these topics in Chapter 4.

1.2.1 Fundamental questions for low-dimensional models

From the standpoint of applications, the first question to address is that of designing the right model for any given problem, i.e. of choosing the right set of equations and boundary conditions for some set of loadings on an elastic object of given properties. This is of course a problem pervasive to all of mathematical modeling, but it is certainly acute for plate theories, where choices abound and application domains have to be determined with complicated heuristics: *how thin is this plate, what kinds of loads is it subjected to, what are the maximal deformations expected, can we assume that the strain-stress law is linear*, etc.[1.8]

There are essentially two methods to arrive at lower dimensional theories. At the core of both classical and current engineering approaches is the technique of making principled a priori kinematical assumptions defining the structure of admissible displacement and stress fields. Through great physical intuition, theories like e.g. Bernoulli and Timoshenko beams, (linear / nonlinear) Kirchhoff-Love, Reissner and Midlin and von Kármán plate theories were developed and have been in use for over a century.

The other approach, perhaps more natural from a mathematician's point of view is to derive the theories from the classical equations of continuum mechanics. Within this mindset two classical techniqes for plates have been used [Cia97]:

- Direct estimation of the difference between 3D solutions and some given 2D solution by means of embeddings or restrictions. This was done in the context of linear elasticity around the 1950-1970s.

1.8. Furthermore, even when the models are assumed given and one must only choose, many of the same questions arise, for instance if determining the validity of linear elastic approximations (as opposed to nonlinear or elastoplastic). For just one example of this in the context of linearly elastic plate theories, see [AMZ02].

- Formal asymptotic method: starting from an Ansatz based on physical intuition or existing theories in engineering, a (formal) series expansion of 3D displacements in terms of the (dimensionless) thickness of the plate h is made. Higher order terms are discarded and h is sent to 0. Then convergence of the "leading term of the expansion" $u_h \to u$ is proved.

An obvious mathematical question is that of rigorous justification of models obtained in such ways from first principles. Typically this means starting with the most general variational principle possible (minimisation of the stored energy density of a general hyperelastic material under physically realistic conditions) and arriving at the lower dimensional theories by some notion of variational convergence.[1.9] This is the path followed in the recent literature and in this work.

1.3 Justifying lower dimensional theories

At a high level, the task can be expressed as follows:

Algorithm (big picture)

Given an approximate, lower dimensional problem P_0 construct a sequence of problems $(P_h)_{h\to 0}$, converging to P_0, such that the solutions u_h of P_h converge to the solution u_0 of P_0.

One must of course define in which sense these objects converge or are close to each other, and work around the difficulty that the problems P_h may lack solutions.

It is because of this broad scheme that the notion of Γ-convergence is so useful. Roughly speaking, if P_h and P_0 are (re-)written as minimisation problems, Γ-convergence of P_h to P_0 implies convergence of (approximate) minimisers u_h to the miminiser u_0. Note however that, if P_0 is an approximation to some "real" problem P_r, this method does not provide

---

1.9. Note that this considers the classical equations of continuum mechanics as first principle. Another, perhaps more "fundamental", approach is to descend to atomic interactions and their structural arrangements under given potentials. From this discrete setting limits are computed which directly lead to many continuous theories. For a linear example see [Sch09].

any fine estimates on the proximity of u_0 to u_r (or of P_0 to P_r for that matter).[1.10]

We now set some basic nomenclature and notation and review relevant prior art and how it relates to our contributions. The literature on the derivation of effective theories via Γ-convergence is vast, so we will focus on a few cornerstone papers related to plate theories, while briefly mentioniong related ones.

1.3.1 Previous work

A **plate** is a three-dimensional elastic body with two special geometric features: **flatness** (the middle layer of the body is a plane) and **thinness** (one of its dimensions is "much smaller" than the other two).[1.11] Because these features are pervasive in engineering (e.g. in roofs, ship decks and bridges to cite a few applications), it is of great practical interest to learn how these bodies behave under different types of loads and conditions.[1.12]

If external loads act exclusively on and along the midplane, one talks of **plane stress**: the stresses and strains remain planar and are uniformly distributed. When the strain / stress relation remains linear under the loads considered, so called **membrane models** are applicable. If however, loads are transversal to the midplane, in particular normal to it, the strains and stresses cease to be uniform across the midplane and so called **bending phenomena** become relevant. The resulting bending can occur without extension, i.e. no stretching or contraction of the midplane (**pure bending**) or with it (**membrane bending** or **shell-like** behaviour). An inmediate step further is to consider both in-plane and out-of-plane loads, leading to mixed membrane and bending behaviour, present e.g. in von Kármán models, which we will focus upon in the coming chapters.

1.10. Other than the following "trivial" one: if the real problem P_r is included in the sequence $(P_h)_{h\to 0}$, i.e. $r = h^{(r)}$ for some $h^{(r)} \ll 1$, then $\|u_r - u_0\| = \|u_{h^{(r)}} - u_0\| < \varepsilon$ if $h^{(r)}$ is small enough. [P15] suggest doing this systematically for the design of non-standard sequences P_h yielding both common and novel limit models. Their proposal highlights the fact that Γ-convergence results are mathematically rigorous ways of obtaining a particular set of equations from another, which do not show either of them to be physically sound.

1.11. But not too much: extremely thin materials, like fabrics, are not modelled by thin plate models.

1.12. As already mentioned, in this work we focus on the elastic regime for multilayered plates, leaving aside plastic, viscoelastic or any other effects.

A domain $\Omega_h = \omega \times (-h/2, h/2) \subset \mathbb{R}^3$, the **physical plate**, is identified with a hyperelastic body of height h "much smaller" than the lengths of the sides of ω.[1.13] The plane domain $\omega \times \{0\} \subset \mathbb{R}^2$ constitutes the **mid-layer** of the plate. In order to avoid working on a changing domain, a rescaling $x_3 = z_3/h$ is performed to obtain a fixed Ω_1. We set $z_h(x_1, x_2, x_3) = (x_1, x_2, h\, x_3)$ and we consider instead of a deformation $\tilde{y}: \Omega_h \to \mathbb{R}^3$, the **rescaled** one $y_h: \Omega_1 \to \mathbb{R}^3$, $y_h(x) = \tilde{y}(z_h(x))$. We assume that the body has a (possibly non-homogeneous) stored energy density W (precise conditions on W will be specified later) and total **elastic energy** given by $E_h(\tilde{y}) = \int_{\Omega_h} W(z, \nabla \tilde{y}(z))\, dz$. We define the **energy per unit volume** as $J_h = \frac{1}{h} E_h$, which after a change of variables can be seen to be

$$J_h(y) = \int_{\Omega_1} W(x, \nabla_h y)\, dx,$$

where $\nabla_h = (\partial_1, \partial_2, \partial_3/h)$.[1.14] We are interested in minimal energy deformations for J_h and their properties. The goal is to obtain a functional in the Γ-limit $h \to 0$, taking functions of $x' = (x_1, x_2)$ as input, whose minimisers solve the equations of known or novel models. We will not be considering body forces for simplicity, but including them in the analysis as in [FJM06] is straightforward.

1.3.1.1 Linear models

One of the first applications of Γ-convergence to derive limit theories in linear elasticity was [ABP88], where the authors arrive at theories for linearly elastic plates embedded in elastic bodies under a range of scalings of the plate's energy. However, because they assume convex stored energy densities (and therefore not frame indifferent) and consider energies particular to the embedding problem with an additional term for the surrounding body, they do not recover the classical Kirchhoff-Love limits nor strong convergence of solutions [Cia97, §1.11].

1.13. Typical values here are $h = 10^{-2}$ or $h = 10^{-3}$, depending on the application.

1.14. One computes first $\nabla_x y_h(x) = \nabla_z \tilde{y}(z_h(x))\, \nabla_x z_h(x)$ and rearranges to obtain $\nabla_z \tilde{y}(z_h(x)) = \nabla_x y_h(x) (\nabla_x z_h(x))^{-1} = (\partial_1, \partial_2, h^{-1}\partial_3) y_h(x) = \nabla_h y(x)$. Then $E_h(\tilde{y}) = \int_{\Omega_h} W(z, \nabla_z \tilde{y}(z_h(x)))\, |J z_h(x)|\, dx = h \int_{\Omega_1} W(x, \nabla_h y_h(x))\, dx = h\, J_h(y_h)$.

Some of those issues were addressed later in [ABP94]. Jumping to more recent results in the linearly elastic field, we find e.g. justifications of Timoshenko's beam theory [FP15], of Reissner-Mindling plates [PT07] and of some polymer gel models [PT17].

Finally, although not an example of a lower dimensional theory, [DNP02] is interesting for obtaining 3D linear elasticity from the non linear case, using the geometric rigidity result from [FJM02] (see below). This is done under a strong coercivity assumption ($p = 2$) which is relaxed later in [ADD12] to a setting relevant in more applications ($p = 2$ close to SO(3) and $1 < p \leqslant 2$ far from it).

1.3.1.2 Nonlinear models

For the case of strings, the first work to derive a non linearly elastic, lower dimensional theory with an analysis using variational convergence was possibly [ABP91]. In the context of nonlinear plates, it is perhaps most ilustrative to analyze past developments at each scaling of the rescaled functionals (recall that we have rescaled the domain so we are considering energy per unit volume $J_h = \frac{1}{h} E_h$)

$$J_h^{\beta}(y) = \frac{1}{h^{\beta}} \int_{\Omega_1} W(\nabla_h y).$$

$\beta = 0$: [LR95] is the first derivation of a nonlinear plate theory, inspired on the work in [ABP91]: a **large deformation**,[1.15] **frame-indifferent**,[1.16] **non-linear membrane theory**, whose characteristic property is that "non-linear membranes offer no resistance to **crumpling**" (concentration of energy at singular regions due to the confinement to small domains). This

---

1.15. After de-scaling deformations are $\mathcal{O}(1)$ of h.

1.16. Note that, as we do later, they define frame indifference as the property that for every F, $W(RF) = W(F)$ holds for all $R \in$ SO(3), the group of real orthogonal matrices with positive determinant. We do not allow for the weaker condition $R \in O(3)$, since models invariant under this group violate the lower bound $W(F) \geqslant C \operatorname{dist}^2(F, \text{SO}(3))$ (take $F = -I$).

is proved by characterising a subset of deformations corresponding to crumpling in the kernel of the limit stored energy function.

$\beta \in (0, 1)$: This is the so-called **constrained membrane** regime, analysed in detail in [Con04].

$\beta \in [1, 2)$: To the best of our knowledge, this regime remains not very well explored, except under certain kinds of boundary conditions or assumed admissible deformations, exactly as already stated in [FJM06] back in 2006. For instance $\beta = 1$ is the scaling for thin sheets under compressive Dirichlet conditions [BCDM02]. Following some investigation in the physics literature of characteristic modes of crumpling, [CM08] find the Γ-limit for $\beta \in (0, 5/3)$ under confinement to be trivial using approximations by piecewise affine, isometric maps (modeling folding, or "origami"). For the limit case $\beta = 5/3$ which is conjectured to be the proper scaling for crumpling under confinement, they prove an upper bound as well as a lower bound, albeit the latter for specific maps.

$\beta = 2$: [FJM02] prove the fundamental **geometric rigidity** estimate which carries Korn's inequality to the nonlinear setting and utilize it to obtain the non-linear Kirchhoff theory of pure bending under an isometry constraint. See Appendix A.3 for the details. This estimate is at the core of most of the later developments in this area.

$\beta \in [2, \infty)$: In their seminal paper [FJM06], the authors exploit the quantitative geometric rigidity estimate of [FJM02] in a systematic investigation of limits for the whole range of scalings $\beta \in [2, \infty)$, deriving the first **hierarchy** of limit models. They also provide a thorough (albeit succinct) overview of the state of the art around 2006. The lecture [DFMŠ17, Chapter 2] by Müller, provides a nice waltkthrough of this paper, as well as abundant references and open problems as of 2017.

1.3.1.3 Prestrained models

The focus in this thesis is on materials whose reference configuration is subjected to stresses (one speaks of **prestrained** or **prestressed** bodies) and whose energy density exhibits a dependence on the out-of-plane direction (modeling **multilayered** plates). Examples of these situations are heated materials, crystallisations on top of a substrate and multilayered plates. Some of the previous work, again sorted by scaling regime follows:

$\beta = 2$: The study of prestrained plates in this regime begins in [Sch07a] using the techniques in [FJM06, Pak04]. The essential Ansätze are that the stored energies are given either as

$$W(x_3, F) = W_0(a_0(x_3)^{-1} F)$$

or as

$$W(x_3, F) = W_0((1 + h^\alpha f(x_3/h))^{-1} F)$$

with $\alpha = 1$ (the full range $\alpha > 0$ with adequate energy scalings is partially studied in [Ves12]), which can be thought to model respectively thermal stress on a single material and stresses in multilayered films due to mismatching energy wells but homogeneously changing elastic constants: notice that the second energy is minimal whenever $F \in (1 + h^\alpha f(x_3))$ SO(3) and W_0 does not depend on the out-of-plane coordinate so that the elastic constants only experience first order changes (the energy well SO(3) is shifted by vanishing perturbations of the identity). Because of this, the model does not allow for different materials stacked on top of one another, since that requires abrupt changes to the energy wells. One can see how minimisers are modified in the Γ-limit

$$I^0(y) = 1_{\mathscr{A}}(y) \frac{1}{24} \int_S Q_2(\mathrm{II} - a_1 \mathrm{Id}) - a_2 \, dx,$$

where

$$1_{\mathscr{A}}(y) = \begin{cases} 1, & y \in \mathscr{A}, \\ \infty, & \text{else}, \end{cases}$$

and $\mathscr{A}$ is a suitable class of admissible deformations and $a_1 = a_1(f), a_2 = a_2(f)$ can be explicitly computed. This is a "spherical" or isotropic perturbation of the shape tensor II (the second fundamental form of y) in the bending energy $\int Q_2(\mathrm{II})$, which is determined by the x_3 dependency in the argument of W_0.

One application is modeling stationary heating of a single material from below using a monotone f. The top layer would have a smaller factor inducing it to contract, while the lower layer would tend to expand (both with a scaling of h^α). This would induce a curving of the material into a slightly U-shaped figure.

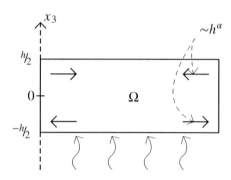

Fig. 1.2. Thermal stresses (decreasing $f(x_3)$) induce displacements of order $\mathcal{O}(h^\alpha)$.

In this situation the reference state $y(x) = x$ has non-vanishing energy and $\nabla y = \mathrm{Id}, \mathrm{II}(y) = 0 \Rightarrow (\ldots) \Rightarrow Q_2(\mathrm{II} - a_1\mathrm{Id}) - a_2 \neq 0$.

[Sch07b] is a natural continuation of this work, with stored energy densities depending on the out-of-plane coordinate $W(x_3, F) = W(x_3, F\,(I + h\,B^h(x_3)))$, thus enabling models of heterogeneous multilayered materials. This is the approach we follow in Chapter 2, although we later leave the heterogeneous setting.

$\beta = 4$: Very closely related to our work in an interpolating theory (see Chapter 2), [LMP11] derive the von Kármán equations for prestressed plates. They assume the domains Ω_h to "undergo a growth process", described by a smooth tensor a_h with $\det a_h(x) > 0$. This tensor is postulated to be one of the factors in a multiplicative decomposition of the deformation gradient: $\nabla y = F\,a_h$ and to have the decomposition $a_h(x', x_3) = \mathrm{Id} + h^2\,\epsilon_g(x') + h\,x_3\,\kappa_g(x')$, with the scalings relevant for the von Kármán regime: h^2 for in-plane and h for out-of-plane displacements. The stored energy density $W(\cdot)$ is then taken to depend only on $F = \nabla y\,(a_h)^{-1}$.

Note that this generalizes our setting in the case $\alpha = 3$, $\beta = 4$ for a particular form of dependence on x_3 since we take $F = \nabla y\, Z^h$, with $Z^h = I + h^{\alpha-1}\, B^h(x_3)$, $\|B^h\|_\infty \leqslant C$ and for h small enough there exists $a_h := (Z^h)^{-1}$. Considering a more general dependence on x_3 introduces additional technical difficulties, as will be seen below.

1.3.2 A remark on shell theories

Although we will always work with flat reference configurations, it is noteworthy that the prestressed reference state that we consider, which is due to an energy well for the stored energy density which shifts with the out-of-plane coordinate, can be considered in the framework of non-euclidean reference metrics for **shells**.

Shell models assume a reference state with curved geometry. They have attracted enormous attention and as a matter of fact, most of the program described above has been carried through for them as well. A complete reference in the spirit of [FJM06], is [LP09], which gathers known results for $\beta \geqslant 2$. Closer to our interpolating theory, [LMP14] continue the work in [LMP11] that we mentioned above and obtain a family of theories for residually strained thin shells parametrized by an additional "shallowness parameter" also in the von Kármán regime.

1.4 Outline

In Chapter 2 we present the main contribution of this thesis, which is an extension of [Sch07b] to the whole hierarchy of models for $\beta \geqslant 2$, built upon the core results in [FJM06]. There we also prove the existence of an intermediate von Kármán regime at $\beta = 4$ parametrised by a scalar $\theta \geqslant 0$ interpolating the neighbouring ones $\beta < 4$ (linearised Kirchhoff) and $\beta > 4$ (linearised von Kármán). In the simplest case of x_3-independent stored energy density and linear internal mismatch, the functional obtained is (see Appendix B for notation)

$$\mathscr{I}_{vK}^\theta(u,v) = \frac{\theta}{2}\int_\omega Q_2\left(\nabla_s u + \tfrac{1}{2}\nabla v \otimes \nabla v\right) + \frac{1}{24}\int_\omega Q_2(\nabla^2 v - I), \qquad (\star)$$

where Q_2 is the quadratic form of linear elasticity.[1.17] As can be seen, this incorporates both membrane and bending energy terms so that one expects the parameter θ to "switch" between minimising configurations for one and the other at some critical value $\theta_c > 0$. In Chapter 3 we characterise minimisers for $\beta < 4$ and $\beta > 4$ and are able to prove uniqueness in the case $\beta = 4$ under the assumption that we are "close" ($\theta \ll 1$) to the linearised regime $\beta > 4$.

Finally, Chapter 4 seeks to provide some empirical evidence of this switching behaviour via numerical experiments. We implement a non-conforming discretisation with penalty and prove its Γ-convergence to the continuous interpolating energy ($\star$). To compute (local) minimisers we use a standard discrete gradient flow [Bar15].

1.5 Acknowledgements

It is hard to show due gratitude without being cheesy, bear with me in these few paragraphs.

First and foremost, I am deeply indebted to the lower and upper bounds, who have provided floor and roof all this time. They kept me warm and cozy during the winter days and cool during summer. Also, to the mollifier, for being so smooth. Only with its constant patience (equal to 1) and increasing focus has it been possible to flatten the wrinkles in this work.

Then of course, I owe my advisor a great debt of gratitude: for his unrelenting confidence since he welcomed me in his department, for proposing such an interesting problem, for his patient guidance and the many hours he has spent holding my mathematical hand through this tortuous path, thank you. This is a debt that I will never be able to repay. So I won't try, ha!

Additionally, many of the hard moments and intermitent feelings of despondency dissipated thanks to the cheerfulness and healthy outlook on life of my colleagues, to whom I am grateful for so many good moments and such an enjoyable workplace.

But it has been the unyielding support, the companionship and intellectual crutches of Ana, my partner in life and everything which have kept me on track, no matter what.

1.17. For instance in an isotropic material this can be the Frobenius norm for adequately chosen Lamé constants.

This work was financially supported by project 285722765 of the DFG, *"Effektive Theorien und Energie minimierende Konfigurationen für heterogene Schichten"*.

Last but not least: this document was entirely prepared using the friction-free, high quality scientific editor for the 21st century, $\text{T}_{\text{E}}\text{X}_{\text{MACS}}$. Decades of man-hours have been invested in making it the best tool available for mathematical editing, and for that I am grateful to all of its developers. Use it now, or regret it later!

2

A hierarchy of multilayered plate models

The focus of this section is the derivation via Γ-convergence of a complete hierarchy of effective plate models with a *prestrained* ground state and inhomogeneous energy density, for scalings ranging from the nonlinear Kirchhoff regime to linearised von Kármán. As mentioned in the introduction, with some simplifications the prototypical limit functional is of von Kármán type:

$$\mathscr{J}_{vK}^{\theta}(u,v) = \frac{\theta}{2} \int_{\omega} Q_2(\nabla_s u + \frac{1}{2}\nabla v \otimes \nabla v)\,dx + \frac{1}{24}\int_{\omega} Q_2(\nabla^2 v - I)\,dx. \qquad (\star)$$

In Section 2.2 we present our main results. Proofs of lower and upper bounds are collected in Section 2.3, where we obtain $(\star)$ and more general functionals. In Section 2.4 we show how the von Kármán functional interpolates between different theories: in the limit $\theta \to \infty$ one obtains a functional with only bending energy and Kirchhoff's isometry constraint, whereas in the limit $\theta \to 0$ the limit is an unconstrained, linear version of von Kármán theory. Finally, in Section 2.5 we prove some density and matrix representation theorems essential for the construction of recovery sequences and identification of minimisers in the linearised Kirchhoff regime.

2.1 The setting

Note. See Appendix B for **notation**.

As described in Section 1.3.1, we consider a sequence of increasingly thin domains $\Omega_h := \omega \times (-h/2, h/2) \in \mathbb{R}^3$ and rescale them to

$$\Omega_1 := \omega \times (-1/2, 1/2) \subset \mathbb{R}^3$$

where ω fulfills:

Assumptions 2.1. *The domain* $\omega \subset \mathbb{R}^2$ *is bounded with Lipschitz boundary.*

As a consequence of the rescaling, instead of maps $y: \Omega_h \to \mathbb{R}^3$, we consider the **rescaled deformations**

$$y_h: \Omega_1 \to \mathbb{R}^3, x \mapsto y_h(x) = y(x_1, x_2, hx_3),$$

belonging to the space

$$Y := W^{1,2}(\Omega_1; \mathbb{R}^3).$$

For each **scaling**[2.1]

$$\alpha \in (2, \infty),$$

and for all deformations $y \in Y$, define the **scaled elastic energy** per unit volume[2.2]:

$$\mathscr{I}_\alpha^h(y) = \frac{1}{h^{2\alpha-2}} \int_{\Omega_1} W_\alpha^h(x_3, \nabla_h y(x)) \, dx, \qquad (2.1)$$

2.1. In the notation of Chapter 1 we have $\beta = 2\alpha - 2$.

2.2. If we start with the energy per unit volume

$$J_\alpha^h(y^h) := \frac{1}{h} \int_{\Omega_h} W_\alpha^h(z_3, \nabla y^h(z)) \, dz,$$

then we have defined, after a coordinate transformation $x_3 = z_3/h$ in the integral:

$$\mathscr{I}_\alpha^h(y) = \frac{1}{h^{2\alpha-2}} E_\alpha^h(y^h).$$

where $\nabla_h = (\partial_1, \partial_2, \partial_3 / h)^\top$ is the gradient operator resulting after the change of coordinates described in Section 1.3.1. For the sake of conciseness, we will present most results below for all scalings simultaneously, adding the parameter α to much of the notation. The energy density for $\alpha \neq 3$ is given by

$$W_\alpha^h(x_3, F) = W_0(x_3, F(I + h^{\alpha-1} B^h(x_3))), \quad F \in \mathbb{R}^{3 \times 3}. \qquad (2.2)$$

where $B^h : \left(-\tfrac{1}{2}, \tfrac{1}{2}\right) \to \mathbb{R}^{3 \times 3}$ describes the **internal misfit** and W_0 the stored energy density of the reference configuration. In the regime $\alpha = 3$ we include an additional parameter $\theta > 0$ controlling further the amount of misfit in the model:

$$W_{\alpha=3}^h(x_3, F) = W_0\left(x_3, F\left(I + h^2 \sqrt{\theta}\, B^h(x_3)\right)\right), \quad F \in \mathbb{R}^{3 \times 3},$$

and we later write $\tilde{B}^h = \sqrt{\theta}\, B^h$. Note that given the choice $h^{\alpha-1}$ for the scaling of the misfit, the fact that in the limit it will be again scaled quadratically forces the choice of a scaling of $h^{-2(\alpha-1)}$ for the energy, since otherwise one would compute trivial (vanishing or infinite) energies in the limits. This will become apparent in the computation of the lower bounds in Theorem 2.8. Our assumptions for B^h and W_0 are those of [Sch07b, Assumption 1.1]:

Assumptions 2.2.

a) For a.e. $t \in \left(-\tfrac{1}{2}, \tfrac{1}{2}\right)$, $W_0(t, \cdot)$ is continuous on $\mathbb{R}^{3 \times 3}$ and C^2 in a neighbourhood of $SO(3)$ which does not depend on t.
b) The quadratic form $Q_3(t, \cdot) = D^2 W_0(t, I)[\cdot, \cdot]$ is in $L^\infty\left(\left(-\tfrac{1}{2}, \tfrac{1}{2}\right); \mathbb{R}^{9 \times 9}\right)$.
c) The map

$$\omega(s) := \operatorname*{ess\,sup}_{-\frac{1}{2} < t < \frac{1}{2}} \sup_{|F| \leqslant s} |W_0(t, I + F) - \tfrac{1}{2} Q_3(t, F)|$$

shall satisfy $\omega(s) = o(s^2)$ as $s \to 0$.
d) For all $F \in \mathbb{R}^{3 \times 3}$ and all $R \in SO(3)$

$$W_0(t, F) = W_0(t, RF).$$

e) For a.e. $t \in \left(-\frac{1}{2}, \frac{1}{2}\right)$, $W_0(t, F) = 0$ if $F \in SO(3)$ and

$$\underset{-\frac{1}{2} < t < \frac{1}{2}}{\text{ess inf}} W_0(t, F) \geqslant C \operatorname{dist}^2(F, SO(3)),$$

for all $F \in \mathbb{R}^{3 \times 3}$ and some $C > 0$.

f) $B^h \to B$ in $L^\infty\left(\left(-\frac{1}{2}, \frac{1}{2}\right); \mathbb{R}^{3 \times 3}\right)$.

We note in passing two consequences of these conditions. First, frame invariance (Assumption 2.2.d) extends to the second derivative where defined, i.e.[2.3]

$$D^2 W_0(t, R)[RF, RF] = D^2 W_0(t, I)[F, F] = Q_3(t, F).$$

Second, the energy W_0 grows at most quadratically in a neighbourhood of $SO(3)$, i.e. for small $|F|$ it holds that:[2.4]

$$W_0(t, I + F) \leqslant C \operatorname{dist}^2(I + F, SO(3)). \tag{2.3}$$

2.3. To see this, let $\varepsilon > 0$ so that the following derivatives are defined and compute, using Taylor as in Footnote 2.4:

$$\frac{\mathrm{d}^2}{\mathrm{d}\varepsilon^2}\Big|_{\varepsilon=0} W_0(t, I + \varepsilon F) = D^2 W_0(t, I)[F, F],$$

which by assumption is equal to:

$$\frac{\mathrm{d}^2}{\mathrm{d}\varepsilon^2}\Big|_{\varepsilon=0} W_0(t, R(I + \varepsilon F)) = D^2 W_0(t, R)[RF, RF].$$

2.4. Let $G = I + F$ and let P be its projection onto the rotations $P = P_{SO(3)} G$. Then we have $\operatorname{dist}^2(I + F, SO(3)) = |G - P|^2$. If $|F|$ is small enough for $I + F$ to be in the set where W_0 is twice differentiable (Assumption 2.2.a) we can consider the Taylor expansion

$$\begin{aligned} W_0(t, G) &= W_0(t, P) + D W_0(t, P)[G - P] + \tfrac{1}{2} D^2 W_0(t, P)[G - P, G - P] + o(|G - P|^2) \\ &= \tfrac{1}{2} D^2 W_0(t, P)[G - P, G - P] + o(|G - P|^2), \end{aligned}$$

where the first and second terms vanish because of Assumption 2.2.e. Now, by frame indifference we have $D^2 W_0(t, P)[G - P, G - P] = Q_3(t, (G - P) P^{-1})$, and by Assumption 2.2.b, we have for a.e. t the bound $Q_3(t, A P^{-1}) \leqslant C |A P^{-1}|^2 = C |A|^2$, where the last equality follows from P^{-1} being orthogonal (Lemma A.3). But then

$$W_0(t, I + F) \leqslant C |G - P|^2 + o(|G - P|^2) \leqslant C \operatorname{dist}^2(I + F, SO(3)).$$

The Hessian

$$Q_3(t, F) := D^2 W_0(t, I)[F, F] = \frac{\partial^2 W_0(t, I)}{\partial F_{ij} \partial F_{ij}} F_{ij} F_{ij},$$

for $t \in \left(-\frac{1}{2}, \frac{1}{2}\right), F \in \mathbb{R}^{3 \times 3}$ is twice the quadratic form of linear elasticity theory, which results after a linearisation of W_0 around the identity.[2.5] Define Q_2 to be the quadratic form on $\mathbb{R}^{2 \times 2}$ obtained by relaxation of Q_3 among stretches in the x_3 direction:

$$Q_2(t, G) := \min_{c \in \mathbb{R}^3} Q_3(t, \hat{G} + c \otimes e_3), \text{ for } t \in \left(-\frac{1}{2}, \frac{1}{2}\right), G \in \mathbb{R}^{2 \times 2},$$

where $e_3 = (0, 0, 1) \in \mathbb{R}^3$. This process effectively minimises away the effect of transversal strain.[2.6] The maps $Q_2(t, \cdot)$ are positive semidefinite, convex and vanish on antisymmetric matrices (Lemma A.13) and there exists a map $\mathscr{L}: I \times \mathbb{R}^{2 \times 2} \to \mathbb{R}^3$, linear in its second argument, which attains the minimum (Lemma A.12):

$$Q_2(t, G) = Q_3(t, \hat{G} + \mathscr{L}(t, G) \otimes e_3).$$

For the regimes $\alpha \geqslant 3$, we define the effective form

$$\overline{Q}_2(E, F) := \int_{-\frac{1}{2}}^{\frac{1}{2}} Q_2(t, E + t F + \check{B}(t)) \, dt, \tag{2.4}$$

with $E, F \in \mathbb{R}^{2 \times 2}$ and for $\alpha \in (2, 3)$ its relaxation

$$\overline{Q}_2^\star(F) := \min_{E \in \mathbb{R}^{2 \times 2}} \overline{Q}_2(E, F) = \min_{E \in \mathbb{R}_{\text{sym}}^{2 \times 2}} \int_{-\frac{1}{2}}^{\frac{1}{2}} Q_2(t, E + t F + \check{B}(t)) \, dt. \tag{2.5}$$

For the case $\alpha = 3$, we include an additional parameter $\theta > 0$ as discussed in page 27 and later write $\tilde{B} = \sqrt{\theta} B$.

2.5. As explained in Section 1.1, for geometrically linear materials its argument F would be the linear strain tensor $e(w) := \nabla_s w$.

2.6. Indeed, with $F = \nabla u$ the components minimised away are $(F_{13}, F_{23}, F_{33}) = (\partial_3 u_1, \partial_3 u_2, \partial_3 u_3)$.

Remark 2.3. Note that $\overline{Q}_2$ and $\overline{Q}_2^\star$ are convex polynomials of degree 2 and $\overline{Q}_2(E,F), \overline{Q}_2^\star(F)$ are in $L^1(\omega)$ for all $E, F \in L^p(\omega; \mathbb{R}^{2\times 2})$, $p \geqslant 2$.

For fixed $\alpha \in (2, \infty)$ we say that a sequence $(y^h)_{h>0} \subset Y$ has **finite scaled energy** if there exists some constant $C > 0$ such that

$$\operatorname*{lsup}_{h\to 0} \mathscr{I}_\alpha^h(y^h) \leqslant C. \tag{2.6}$$

This definition will be central for many of the arguments below. After some corrections we will have precompactness of such sequences, thus proving that the family $\mathscr{I}_\alpha^h$ is equicoercive, the essential condition for the fundamental theorem of Γ-convergence showing convergence of minimisers and energies, cf. [Bra06] and Section A.5. This compactness takes place in adequate **target ambient spaces**

$$X_\alpha = \begin{cases} W^{1,2}(\omega; \mathbb{R}) & \text{if } \alpha \in (2,3), \\ W^{1,2}(\omega; \mathbb{R}^2) \times W^{1,2}(\omega; \mathbb{R}) & \text{if } \alpha \geqslant 3, \end{cases}$$

equipped with the **weak topology**.[2.7]

An essential ingredient in arguments with Γ-convergence is the choice of sequential convergence to obtain (pre-)compactness. By Remark A.25, for the lower bounds we may suppose that a sequence $(y^h)_{h>0}$ has finite scaled energy, which enables Lemma A.30 for the identification of the limits. We choose to encode the necessary estimate for the gradient of the deformations into the definition of *convergence via maps P_α^h* (Definition 2.4). Despite adding clutter to the notation, this helps to highlight and isolate the technical requirement of using sequences which are close to the identity, up to certain rigid transformations.

Definition 2.4. *Let* $Y := W^{1,2}(\Omega_1; \mathbb{R}^3)$ *and*

$$X_\alpha := \begin{cases} W^{1,2}(\omega; \mathbb{R}) & \text{if } \alpha \in (2,3), \\ W^{1,2}(\omega; \mathbb{R}^2) \times W^{1,2}(\omega; \mathbb{R}) & \text{if } \alpha \geqslant 3. \end{cases}$$

2.7. Because the weak topology is not 1st countable, for Γ-convergence one argues that one may consider bounded sets, where it is metrisable.

We say that a sequence $(y^h)_{h>0} \subset Y$ $\boldsymbol{P^h}$-converges to some $w \in X_\alpha$ if and only if there exist maps $R^h \colon \Omega_1 \to \mathrm{SO}(3)$, constant along x_3, such that

$$\|\nabla_h y^h - R^h\|_{0,2,\Omega_1} \leqslant C h^{\alpha-1} \text{ with } \|R^h - I\|_{0,2,\Omega_1} \leqslant C h^{\alpha-2},$$

and

$$P_\alpha^h(y^h) \to w \quad \text{weakly in } X_\alpha,$$

where

$$P_\alpha^h \colon Y \to X_\alpha, y^h \mapsto \begin{cases} v_\alpha^h, & \text{if } \alpha \in (2,3), \\ (u_\theta^h, v_\theta^h) & \text{if } \alpha = 3, \\ (u_\alpha^h, v_\alpha^h), & \text{if } \alpha > 3, \end{cases}$$

and we defined:

*For $\alpha \neq 3$ and $x' \in \omega$, the **scaled and averaged in-plane** and **out-of-plane** displacements:*

$$\begin{cases} u_\alpha^h(x') := \dfrac{1}{h^\gamma} \displaystyle\int_{-1/2}^{1/2} (y^{h'}(x',x_3) - x') \, dx_3, \\[2mm] v_\alpha^h(x') := \dfrac{1}{h^{\alpha-2}} \displaystyle\int_{-1/2}^{1/2} y_3^h(x',x_3) \, dx_3, \end{cases} \qquad (2.7)$$

where

$$\gamma = \begin{cases} 2(\alpha-2) & \text{if } \alpha \in (2,3), \\ \alpha-1 & \text{if } \alpha > 3. \end{cases}$$

For $\alpha = 3$ and $x' \in \omega$, we introduce the additional parameter $\theta > 0$:

$$\begin{cases} u_\theta^h(x') := \dfrac{1}{\theta h^2} \displaystyle\int_{-1/2}^{1/2} [y^{h'}(x',x_3) - x'] \, dx_3 \\[2mm] v_\theta^h(x') := \dfrac{1}{\sqrt{\theta}\, h} \displaystyle\int_{-1/2}^{1/2} y_3^h(x',x_3) \, dx_3. \end{cases} \qquad (2.8)$$

For $\alpha = 3$, *we overload the notation with the parameter θ writing (u_θ^h, v_θ^h) and P_θ^h instead of (u_α^h, v_α^h) or P_α^h, letting the letter used in the subindex resolve ambiguity.*

With Definition 2.4 we can specify precisely what we mean by Γ-convergence of the energies (2.1):[2.8]

Definition 2.5. *Let $\alpha > 2$. We say that the family of scaled elastic energies $\{\mathscr{I}_\alpha^h : Y \to \mathbb{R}\}_{h>0}$, $h > 0$, Γ-converges via maps P^h to $\mathscr{I}_\alpha : X_\alpha \to \mathbb{R}$ iff:*

*a) **Lower bound:** For every $w \in X_\alpha$ and every sequence $(y^h)_{h>0} \subset Y$ which P^h-converges to w as $h \to 0$ it holds that*

$$\liminf_{h \to 0} \mathscr{I}_\alpha^h(y^h) \geqslant \mathscr{I}_\alpha(w).$$

*b) **Upper bound:** For every $w \in X$ there exists a **recovery sequence** $(y^h)_{h>0} \subset Y$ which P^h-converges to w as $h \to 0$ and*

$$\limsup_{h \to 0} \mathscr{I}_\alpha^h(y^h) \leqslant \mathscr{I}_\alpha(w).$$

Finally, we identify what the space of **admissible displacements** for the limit theories will be:

$$X_\alpha^0 := \begin{cases} W_{sh}^{2,2}(\omega; \mathbb{R}) & \text{if } \alpha \in (2,3), \\ W^{1,2}(\omega; \mathbb{R}^2) \times W^{2,2}(\omega; \mathbb{R}) & \text{if } \alpha \geqslant 3, \end{cases}$$

where the space of out-of-plane displacements with singular Hessian

$$W_{sh}^{2,2}(\omega) := \{v \in W^{2,2}(\omega; \mathbb{R}) : \det \nabla^2 v = 0 \text{ a.e.}\},$$

will be central in the linearised Kirchhoff theory. We define the functionals to be $+\infty$ for inadmissible displacements in $X_\alpha \backslash X_\alpha^0$.

2.8. We refer to the notes [Bra06] for a quick introduction to Γ-convergence. For a concise collection of the results we require, adapted to this definition, see Appendix A.5.

2.2 Main results

Our first goal is to prove that in the prestrained setting described above one has a hierarchy of plate models *à la* [FJM06]. The proof is split into several theorems in Section 2.3. For notation and details on our particular use of Γ-convergence, see Appendix A.5.

Theorem 2.6. *Let*

$$\mathscr{I}_\alpha^h(y) = \frac{1}{h^{2\alpha-2}} \int_{\Omega_1} W_\alpha^h(x_3, \nabla_h y(x)) \, dx.$$

If $\alpha \in (2,3)$ *and **ω is convex**, then the elastic energies* $\mathscr{I}_\alpha^h$ *Γ-converge to the **linearised Kirchhoff** energy*[2.9]

$$\mathscr{I}_{lKi}(v) := \begin{cases} \frac{1}{2} \int_\omega \overline{Q}_2^\star(-\nabla^2 v) & \text{if } v \in W_{sh}^{2,2}(\omega), \\ \infty & \text{otherwise,} \end{cases} \tag{2.9}$$

where $\overline{Q}_2^\star$ *is defined in (2.5). See Theorems 2.8 and 2.9.*

If $\alpha = 3$ *and* $\theta > 0$ *then the energies* $\mathscr{I}_\theta^h := \frac{1}{\theta} \mathscr{I}_{\alpha=3}^h$ *Γ-converge to the **von Kármán type energy**[2.10]*

$$\mathscr{I}_{vK}^\theta(u,v) := \begin{cases} \frac{1}{2} \int_\omega \overline{Q}_2(\theta^{1/2}(\nabla_s u + \frac{1}{2}\nabla v \otimes \nabla v), -\nabla^2 v) \\ \qquad \text{if } (u,v) \in W^{1,2}(\omega; \mathbb{R}^2) \times W^{2,2}(\omega; \mathbb{R}), \\ \infty, \text{ otherwise,} \end{cases} \tag{2.10}$$

where $\overline{Q}_2$ *is defined in (2.4). See Theorems 2.8 and 2.10.*

Finally, if $\alpha > 3$ *then* $\mathscr{I}_\alpha^h$ *Γ-converges to the **linearised von Kármán** energy*

$$\mathscr{I}_{lvK}(u,v) := \begin{cases} \frac{1}{2} \int_\omega \overline{Q}_2(\nabla_s u, -\nabla^2 v), \\ \qquad \text{if } (u,v) \in W^{1,2}(\omega; \mathbb{R}^2) \times W^{2,2}(\omega; \mathbb{R}) \\ \infty, \text{ otherwise.} \end{cases} \tag{2.11}$$

See Theorems 2.8 and 2.11.

2.9. Convexity of the domain is required for the representation theorems in Section 2.5 which are used in the construction of the recovery sequence for $\alpha \in (2,3)$.

2.10. Again, we slightly overload the notation in what would be a double definition of $\mathscr{I}_3^h$, trusting the letter used in the subindex to dispel the ambiguity.

The functional $\mathcal{J}_{lKi}$ is said to model a **linearised Kirchhof** regime because the isometry condition $\nabla^T y \, \nabla y = I$ of the Kirchhoff model is replaced by $\det \nabla^2 v = 0$, a necessary and sufficient condition for the existence of an in-plane displacement u such that $\nabla u + \nabla^T u + \nabla v \otimes \nabla v = 0$, [FJM06, Proposition 9]. This condition is to leading order equivalent to $\nabla^T y \, \nabla y = I$ for deformations $y = (h^{2\alpha-4} u, h^{\alpha-2} v)$.[2.11] The functional $\mathcal{J}_{vK}^\theta$ is of **von Kármán** type with in-plane and out-of-plane strains interacting in a membrane energy term, and a bending energy term. For simple choices of Q_2 and B^h, one recovers the classical functional ($\star$). Finally, we say that the third limit $\mathcal{J}_{lvK}$, models a **linearised von Kármán** regime by analogy with the classical equivalent, but it is of a different kind than the one expected from the hierarchy derived in [FJM06], since it again features an interplay between in-plane and out-of-plane components.[2.12]

Our second goal is to show that the limit energy $\mathcal{J}_{vK}^\theta$ interpolates between $\mathcal{J}_{lKi}$ and $\mathcal{J}_{lvK}$ as the parameter θ moves from ∞ to 0. In this sense one can say that the theory of von Kármán type bridges the other two. More precisely, in Section 2.4 we prove:

Theorem 2.7. (Interpolating regime) *Assume ω is convex. The following two Γ-limits hold:*

$$\mathcal{J}_{vK}^\theta \xrightarrow[\theta\uparrow\infty]{\Gamma} \mathcal{J}_{lKi},$$

(Theorems 2.15 and 2.16) and

$$\mathcal{J}_{vK}^\theta \xrightarrow[\theta\downarrow 0]{\Gamma} \mathcal{J}_{lvK}$$

2.11. In the numerical analysis literature, the denomination *linear Kirchhoff* is sometimes used for a pure bending regime without constraints.

2.12. This is in contrast to [FJM06]. In our setting with the additional dependence on the x_3 coordinate, it is not possible to simply drop terms while bounding below the energy in the proof of the lower bound as is done in [FJM06, p. 211] because of the difficulty in building recovery sequences later. For $\alpha \in (2,3)$ we introduce an additional relaxation and make use of representation Theorem 2.22 to construct them, but for $\alpha > 3$, no such result is available. One could think that minimising globally, $\inf_{\nabla_s u} \int Q_2(t, \nabla_s u + \ldots)$, might be a way of discarding in-plane displacements to recover the standard theory, but this yields a functional which is not local and therefore lacks an integral representation (see e.g. [BD98, Chapter 9]). Note that even if we pick Q_2 independent of t and $B=0$, we do not recover the functional of [FJM06] because ours keeps track of both in-plane and out-of-plane displacements which is essential to capture the effect of pre-stressing with the internal misfit B^h.

(Theorems 2.17 and 2.18). Furthermore, sequences $(u_\theta, v_\theta)_{\theta>0}$ of bounded energy $\mathscr{I}_{vK}^\theta$ are precompact in suitable spaces as $\theta \uparrow \infty$ or $\theta \downarrow 0$ (Theorem 2.14).

2.2.1 The limit energies for a linear internal mismatch

To show some concrete examples of how the functionals in Theorem 2.6 look like, we choose now a linear internal misfit

$$B(t) := t I_3 \in \mathbb{R}^{3\times3}, \tag{2.12}$$

and consider two different choices for $Q_2(t, \cdot)$.

Separate variables in Q_2: If we assume that $Q_2(t, A) = f(t)\, \tilde{Q}_2(A)$ for some *even* function f. Then, using the definition of $\overline{Q}_2$ from (2.4) we can compute

$$
\begin{aligned}
\overline{Q}_2(E,F) &= \int_{-\frac{1}{2}}^{\frac{1}{2}} f(t)\, \tilde{Q}_2(E + t(F+I))\, dt \\
&= \int_{-\frac{1}{2}}^{\frac{1}{2}} f(t)\, \tilde{Q}_2(E)\, dt + \int_{-\frac{1}{2}}^{\frac{1}{2}} f(t)\, t^2\, \tilde{Q}_2(F+I)\, dt \\
&\quad + \underbrace{\int_{-\frac{1}{2}}^{\frac{1}{2}} 2f(t)\, t\, \tilde{Q}_2[E, F+I]\, dt}_{=0} \\
&= C_1 \tilde{Q}_2(E) + C_2 \tilde{Q}_2(F+I).
\end{aligned}
$$

And the limit energy is then

$$\mathscr{I}_{vK}^\theta(u,v) = \frac{C_1 \theta}{2} \int_\omega \tilde{Q}_2\left(\nabla_s u + \frac{1}{2}\nabla v \otimes \nabla v\right) + \frac{C_2}{2} \int_\omega \tilde{Q}_2(\nabla^2 v - I).$$

One particular instance is having Q_2 *independent of t*, i.e. the **homogeneous case** $\tilde{Q}_2(A) = Q_2(A)$, and $f \equiv 1$. Then

$$\mathscr{I}_{vK}^\theta(u,v) = \frac{\theta}{2} \int_\omega Q_2\left(\nabla_s u + \frac{1}{2}\nabla v \otimes \nabla v\right) + \frac{1}{24} \int_\omega Q_2(\nabla^2 v - I). \tag{2.13}$$

Analogously, in this setting:

$$\mathscr{I}_{lvK}(u,v) = \frac{1}{2} \int_\omega Q_2(\nabla_s u) + \frac{1}{24} \int_\omega Q_2(\nabla^2 v - I).$$

These functionals, where the elastic coefficients do not depend on the out-of-plane component, can model for instance a single-layer material under thermal stress, see Figure 1.2 and the discussion leading to it. We will be mostly studying the energy (2.13) as a function of θ, in particular in Chapters 3 and 4 when we investigate minimal configurations analytically and numerically.

General $Q_2(t,\cdot)$: If we do not assume anything about the dependency of Q_2 on t, an analogous computation with (2.4) shows that the limit energy for $\alpha = 3$ is:

$$
\begin{aligned}
\mathcal{J}_{vK}^{\theta}(u,v) \;=\; & \frac{\theta}{2} \int_{\Omega_1} Q_2(x_3, \nabla_s u + \tfrac{1}{2} \nabla v \otimes \nabla v) \, dx \\
& + \frac{1}{2} \int_{\Omega_1} x_3^2 Q_2(x_3, \nabla^2 v - I) \, dx \\
& - \sqrt{\theta} \int_{\Omega_1} x_3 Q_2 \Big[x_3, \nabla_s u + \tfrac{1}{2} \nabla v \otimes \nabla v, \nabla^2 v - I \Big] \, dx,
\end{aligned}
$$

and for $\alpha > 3$:

$$
\begin{aligned}
\mathcal{J}_{lvK}(u,v) \;=\; & \frac{1}{2} \int_{\Omega_1} Q_2(x_3, \nabla_s u) \, dx + \frac{1}{2} \int_{\Omega_1} x_3^2 Q_2(x_3, \nabla^2 v - I) \, dx \\
& - \int_{\Omega_1} x_3 Q_2[x_3, \nabla_s u, \nabla^2 v - I] \, dx.
\end{aligned}
$$

A particular case of this situation is a bundle of plates modelled by a piecewise constant (wrt. t) form $Q_2(t, A) = \sum_{j=1}^{m} \chi_{[a_{j-1}, a_j)}(t) \, Q_2^j(A)$. By the previous computation the limit energy for $\alpha = 3$ will be

$$
\begin{aligned}
\mathcal{J}_{vK}^{\theta}(u,v) \;=\; & \frac{\theta}{2} \sum_{j=1}^{m} \int_{\omega} (a_j - a_{j-1}) Q_2^j (\nabla_s u + \tfrac{1}{2} \nabla v \otimes \nabla v) \, dx' \\
& + \frac{1}{2} \sum_{j=1}^{m} \int_{\omega} \alpha_j Q_2^j (\nabla^2 v - I) \, dx' \\
& - \sqrt{\theta} \sum_{j=1}^{m} \int_{\omega} \beta_j Q_2^j \Big[\nabla_s u + \tfrac{1}{2} \nabla v \otimes \nabla v, \nabla^2 v - I \Big] \, dx',
\end{aligned}
$$

where $\alpha_j = (a_j^3 - a_{j-1}^3)/3$ and $\beta_j = (a_j^2 - a_{j-1}^2)/2$. For $\alpha > 3$ we obtain an analogous functional with the membrane strain lacking the nonlinearity $\frac{1}{2}\nabla v \otimes \nabla v$. Notice however the third sum of mixed terms. Their presence makes establishing lower bounds difficult without additional assumptions, thus precluding us from showing e.g. uniqueness of minimisers for all $\theta > 0$ (cf. Chapter 3).

2.3 Γ-convergence of the hierarchy

This subsection proves the lower (Theorem 2.8) and upper bounds (Theorems 2.9, 2.10 and 2.11) required for deriving the hierarchy of models in Theorem 2.6.

An important result of [FJM06] is that for small $h > 0$ deformations y^h of finite scaled energy are, up to rigid motions, roughly the trivial map $(x', x_3) \mapsto (x', h\,x_3)$. The factor by which they fail to (almost) be the identity is essential for the linearisation step in the proof below as well as for the identification of the limit strains of weakly convergent sequences of scaled displacements. We must account for these rigid motions if compactness is to be achieved, in particular because deformations might "wander to infinity" without altering the elastic energy. Lemmas A.30 and A.31 gather these ideas more precisely. In particular, the last statement of A.30 provides the required compactness.

Recall that we are always using weak convergence in the spaces X_α.

Theorem 2.8. (Lower bounds) *Let $\alpha \in (2,3)$. If $(y^h)_{h>0} \subset Y$ is a sequence P_α^h-converging to $v \in X_\alpha$, then*

$$\liminf_{h \to 0} \mathscr{I}_\alpha^h(y^h) \geqslant \mathscr{I}_{lKi}(v).$$

Now let $\alpha = 3$. If $(y^h)_{h>0} \subset Y$ is a sequence P_θ^h-converging to $(u,v) \in X_\alpha$, then for all $\theta > 0$

$$\liminf_{h \to 0} \frac{1}{\theta} \mathscr{I}_\alpha^h(y^h) \geqslant \mathscr{I}_{vK}^\theta(u,v).$$

Finally, let $\alpha > 3$. If $(y^h)_{h>0} \subset Y$ is a sequence P_α^h-converging to $(u,v) \in X_\alpha$, then

$$\liminf_{h \to 0} \mathscr{I}_\alpha^h(y^h) \geq \mathscr{I}_{lvK}(u,v).$$

Proof. If $\alpha = 3$, we define $\tilde{B}^h := \sqrt{\theta}\, B^h$ and $\tilde{B} = \sqrt{\theta}\, B$, otherwise $\tilde{B} := B$ and $\tilde{B}^h := B^h$. Following closely the techniques in [FJM02, FJM06, Sch07a, Sch07b] we use a Taylor expansion of the energy around the identity which allows us to cancel or identify its lower order terms. For this we must correct the deformations with their approximation by rotations and work in adequate sets where there is control over higher order terms.

Step 1: *Rewriting of the deformation gradient.*

Let $R^h \colon \omega \to SO(3)$ approximate $\nabla_h y^h$ in $L^2(\Omega_1)$ as in Definition 2.4. The functions

$$G^h := \frac{(R^h)^\top \nabla_h y^h - I}{h^{\alpha-1}}$$

are uniformly bounded in L^2 by invariance of the norm by rotations:

$$\begin{aligned}
\|G^h\|_{0,2,\Omega_1} &= h^{1-\alpha} \|(R^h)^\top \nabla_h y^h - I\|_{0,2,\Omega_1} \\
&= h^{1-\alpha} \|\nabla_h y^h - R^h\|_{0,2,\Omega_1} \leq C.
\end{aligned} \tag{2.14}$$

Now, by the frame invariance of $W^h(x_3, \cdot)$:

$$\begin{aligned}
W^h(x_3, \nabla_h y^h) &= W^h(x_3, (R^h)^\top \nabla_h y^h) \\
&= W_0(x_3, (R^h)^\top \nabla_h y^h (I + h^{\alpha-1} \tilde{B}^h(x_3))) \\
&= W_0(x_3, I + h^{\alpha-1} A^h),
\end{aligned} \tag{2.15}$$

where we have set

$$\begin{aligned}
A^h(x) &:= \frac{(R^h)^\top \nabla_h y^h(x) - I}{h^{\alpha-1}} + (R^h)^\top \nabla_h y^h(x) \tilde{B}^h(x_3) \\
&= G^h + (R^h)^\top \nabla_h y^h \tilde{B}^h.
\end{aligned}$$

Step 2: *Cutoff function.*

We will be expanding $W_0(x_3, I + h^{\alpha-1} A^h)$ around I, but in order to apply the Taylor expansion successfully we need to stay where W_0 is twice differentiable, that is we must control $\mathrm{dist}(I + h^{\alpha-1} A^h, SO(3))$. We achieve this by multiplying with a cutoff function χ^h, defined as the characteristic function of the "good set" $\left\{ x \in \Omega_1 : |G^h| \leqslant h^{-\frac{1}{2}} \right\}$. Here we have:

$$h^{\frac{1}{2}} \gg h^{\alpha - \frac{3}{2}} \geqslant \chi^h |h^{\alpha-1} G^h| = \chi^h |(R^h)^\top \nabla_h y^h - I| = \chi^h |\nabla_h y^h - R^h|,$$

which, because $|R^h| \equiv \sqrt{3}$, implies that $\chi^h |\nabla_h y^h| \leqslant C$. Consequently, since the $\tilde{B}^h$ are uniformly bounded as well:

$$
\begin{aligned}
\chi^h |h^{\alpha-1} A^h| &= \chi^h |h^{\alpha-1} G^h + h^{\alpha-1} (R^h)^\top \nabla_h y^h \tilde{B}^h| \\
&\leqslant \chi^h |h^{\alpha-1} G^h| + \mathcal{O}(h^{\alpha-1}) \\
&= o\left(h^{\frac{1}{2}}\right),
\end{aligned}
\tag{2.16}
$$

and then

$$\mathrm{dist}(I + h^{\alpha-1} \chi^h A^h, SO(3)) \leqslant |I + h^{\alpha-1} \chi^h A^h - I| = o\left(h^{\frac{1}{2}}\right),$$

so in the good sets we may indeed expand around I for small values of h. Now, the sequence $(G^h)_{h>0}$ is bounded in L^2 by (2.14) so we may extract a subsequence converging weakly in L^2 to some $G \in L^2(\Omega_1)$, which we consider from now on without relabelling.[2.13] Furthermore the sequence $(\chi^h)_{h>0}$ is essentially bounded and $\chi^h \to 1$ in measure in Ω_1 (we say that the χ^h *converge boundedly in measure* to 1, see Appendix A.4 for the definition and properties). Indeed $|\{|\chi^h - 1| > \varepsilon\}| = |\{|G^h| > h^{-1/2}\}| \to 0$ as $h \to 0$ because $\|G^h\|_{0,2,\Omega_1} \leqslant C$ uniformly. Consequently (Lemma A.24) we have

$$\chi^h G^h \rightharpoonup G \text{ in } L^2(\Omega_1).$$

Analogously, the sequence $(\chi^h \tilde{B}^h)_{h>0}$ is essentially bounded and converges in measure to $\tilde{B}$ because $|\{|\chi^h \tilde{B}^h - \tilde{B}| > \varepsilon\}| \leqslant |\{|\tilde{B}^h - \tilde{B}| > \varepsilon\}| + |\{\chi^h = 0\} \cap \{|\tilde{B}| > \varepsilon\}| \to 0$. Hence, using again the strong convergence $(R^h)^\top \nabla_h y^h \to I$ in $L^2(\Omega_1)$ (Lemma A.30):

$$(R^h)^\top \nabla_h y^h \chi^h \tilde{B}^h \rightharpoonup \tilde{B} \text{ in } L^2(\Omega_1).$$

2.13. By Remark A.26 this does not affect the lower bound.

So we conclude

$$\chi^h A^h \rightharpoonup A := G + \tilde{B} \text{ in } L^2(\Omega_1). \tag{2.17}$$

Step 3: *Taylor expansion.*

Because $W_0(x_3, \cdot)|_{SO(3)} \equiv 0$, for any fixed x_3 the lower order terms of its Taylor expansion

$$W_0(x_3, I + E) = W_0(x_3, I) + DW_0(x_3, I)[E] + \frac{1}{2}D^2W_0(x_3, I)[E, E] + o(|E|^2)$$

vanish and we have (for small enough h, as explained above)

$$W_0(x_3, I + h^{\alpha-1} \chi^h A^h) = \frac{1}{2}Q_3(x_3, h^{\alpha-1} \chi^h A^h) + \eta^h(x_3, h^{\alpha-1} \chi^h A^h),$$

where $\eta^h(x_3, h^{\alpha-1} \chi^h A^h) = o(h^{2\alpha-2}|\chi^h A^h|^2)$ represents the higher order terms. Defining the uniform bound

$$\omega(s) := \operatorname*{ess\,sup}_{-1 \leqslant 2r \leqslant 1} \sup_{|M| \leqslant s} |\eta^h(r, M)|,$$

we have $\omega(s) = o(s^2)$ by Assumption 2.2.c, and integrating over the rescaled domain Ω_1 we obtain the estimate:

$$\frac{1}{h^{2\alpha-2}} \int_{\Omega_1} W^h(x_3, \nabla_h y^h) \, dx$$

$$\overset{(2.15)}{\geqslant} \frac{1}{h^{2\alpha-2}} \int_{\Omega_1} W^h(x_3, I + \chi^h h^{\alpha-1} A^h) \, dx$$

$$\geqslant \frac{1}{h^{2\alpha-2}} \int_{\Omega_1} \frac{h^{2\alpha-2}}{2} Q_3(x_3, \chi^h A^h) - \omega(|h^{\alpha-1} \chi^h A^h|) \, dx$$

$$= \frac{1}{2} \int_{\Omega_1} Q_3(x_3, \chi^h A^h) - \frac{1}{h^{2\alpha-2}} \int_{\Omega_1} \omega(|h^{\alpha-1} \chi^h A^h|) \, dx. \tag{2.18}$$

Step 4: *The limit inferior.*

In order to pass to the limit, for the first integral on the right hand side of (2.18) we use that Q_3 is positive semidefinite, therefore convex and continuous, and the integral is w.s.l.s.c. [Dac07]. For the second integral we use again Assumption 2.2.c and the fact that $|h^{\alpha-1} \chi^h A^h| \leqslant h^{1/2}$ to obtain the bound (uniform over Ω_1):

$$\frac{\omega(|h^{\alpha-1} \chi^h A^h|)}{|h^{\alpha-1} \chi^h A^h|^2} \leqslant \sup_{|s| \leqslant h^{1/2}} \frac{\omega(s)}{s^2} \longrightarrow 0 \text{ as } h \to 0.$$

But then, because $\chi^h A^h$ converges weakly in L^2, we have $\|\chi^h A^h\|^2_{0,2,\Omega_1} \leqslant C$ and

$$\frac{1}{h^{2\alpha-2}} \int_{\Omega_1} \omega(|h^{\alpha-1} \chi^h A^h|) \, dx = \int_{\Omega_1} \frac{\omega(|h^{\alpha-1} \chi^h A^h|)}{|h^{\alpha-1} \chi^h A^h|^2} \frac{|h^{\alpha-1} \chi^h A^h|^2}{h^{2\alpha-2}} \, dx$$

$$\leqslant \sup_{|s|\leqslant h^{1/2}} \frac{\omega(s)}{s^2} \underbrace{\int_{\Omega_1} |\chi^h A^h|^2 \, dx}_{\text{uniformly bded.}} \longrightarrow 0$$

as $h \to 0$. Taking the liminf at both sides of (2.18) we have:

$$\liminf_{h\to 0} \frac{1}{h^{2\alpha-2}} \int_{\Omega_1} W^h(x_3, \nabla_h y^h) \, dx$$

$$\geqslant \liminf_{h\to 0} \frac{1}{2} \int_{\Omega_1} Q_3(x_3, \chi^h A^h) \, dx$$

$$\quad - \lim_{h\to 0} \frac{1}{h^{2\alpha-2}} \int_{\Omega_1} \omega(|h^{\alpha-1} A^h|) \, dx$$

$$\geqslant \frac{1}{2} \int_{\Omega_1} Q_3(x_3, G + \tilde{B}) \, dx$$

$$\geqslant \frac{1}{2} \int_{\Omega_1} Q_2(x_3, \check{G} + \check{B}) \, dx,$$

where the last estimate follows trivially from the definition of Q_2. **If $\alpha \geqslant 3$**, by Lemma A.31 the limit strain $\check{G}$ has the representation

$$\check{G}(x) = G_0(x') + x_3 G_1(x'),$$

with G_1 and sym G_0 given respectively by (A.14) and (A.15) as:

$$G_1 = \begin{cases} -\sqrt{\theta} \, \nabla^2 v & \text{if } \alpha = 3, \\ -\nabla^2 v & \text{if } \alpha > 3, \end{cases}$$

and

$$\text{sym } G_0 = \begin{cases} \theta\left(\nabla_s u + \frac{1}{2} \nabla v \otimes \nabla v\right) & \text{if } \alpha = 3, \\ \nabla_s u & \text{if } \alpha > 3. \end{cases}$$

We plug both into the last integral and use the fact that $Q_2(x_3, \cdot)$ vanishes on antisymmetric matrices to obtain

$$
\operatorname*{linf}_{h \to 0} \frac{1}{h^{2\alpha-2}} \int_{\Omega_1} W_\alpha^h(x_3, \nabla_h y^h) \, dx
$$

$$
\geqslant \frac{1}{2} \int_{\Omega_1} Q_2(x_3, G_0(x') + x_3 G_1(x') + \breve{B}(x_3)) \, dx
$$

$$
= \frac{1}{2} \int_\omega \overline{Q}_2(\operatorname{sym} G_0, G_1) \, dx'.
$$

In particular, **if $\alpha = 3$**, we have again:

$$
\operatorname*{linf}_{h \to 0} \frac{1}{\theta h^4} \int_{\Omega_1} W_\alpha^h(x_3, \nabla_h y^h) \, dx \geqslant \frac{1}{2\theta} \int_\omega \overline{Q}_2(\operatorname{sym} G_0, G_1) \, dx'
$$

$$
= \frac{1}{2} \int_\omega \overline{Q}_2(\theta^{1/2}(\nabla_s u + \tfrac{1}{2}\nabla v \otimes \nabla v),
$$

$$
-\nabla^2 v) \, dx'.
$$

If $\alpha \in (2, 3)$, then $\operatorname{sym} G_0$ is unknown, so we must further relax the integrand. With the definition of $\overline{Q}_2^\star$ we see that the final integral above is

$$
\frac{1}{2} \int_{\Omega_1} Q_2(x_3, G_0 - x_3 \nabla^2 v + \breve{B}) \, dx \geqslant \frac{1}{2} \int_\omega \overline{Q}_2^\star(-\nabla^2 v) \, dx'. \qquad \square
$$

We proceed now with the computation of the recovery sequences for each of the three regimes discussed. We assume convexity of the domain in order to apply the representation theorems in Section 2.5.

Theorem 2.9. (Upper bound, linearised Kirchhoff regime) *Assume ω is convex, let $\alpha \in (2, 3)$ and $v \in X_\alpha := W^{1,2}(\omega)$. There exists a sequence $(y^h)_{h>0} \subset Y$ which P^h-converges to v such that*

$$
\operatorname*{lsup}_{h \to 0} \mathscr{I}_\alpha^h(y^h) \leqslant \mathscr{I}_{lKi}(v),
$$

with $\mathscr{I}_{lKi}$ defined as in (2.9) by

$$\mathscr{I}_{lKi}(v) := \begin{cases} \frac{1}{2}\int_\omega \overline{Q}_2^\star(-\nabla^2 v(x'))\,\mathrm{d}x' & \text{if } v \in W_{sh}^{2,2}(\omega), \\ \infty & \text{otherwise.} \end{cases}$$

Proof. We set $\varepsilon = h^{\alpha-2}$, so that $h \ll \varepsilon \ll 1$ and $h^2 \ll \varepsilon h \ll 1$.

<u>Step 1</u>: *Setup and recovery sequence.*

The functional $\mathscr{I}_{lKi}$ is strongly continuous on $W_{sh}^{2,2}(\omega)$ by the continuity and 2-growth of $\overline{Q}_2^\star$. By Theorem 2.19 we have a set $\mathscr{V}_0$ of smooth maps with singular Hessian which is $W^{2,2}$-dense in $W_{sh}^{2,2}$, see (2.27) in page 64. Therefore, by the standard Lemma A.27 it is enough to construct here the recovery sequence. Take then a smooth function $v \in \mathscr{V}_0$. Because $\|\nabla v\|_\infty < C$, for ε small enough there exist by [FJM06, Theorem 7] in-plane displacements $u_\varepsilon \in W^{2,2}(\omega;\mathbb{R}^2) \cap W^{2,\infty}(\omega;\mathbb{R}^2)$ with uniform bounds in ε such that the deformations

$$\overline{y}_\varepsilon(x') := \begin{pmatrix} x' + \varepsilon^2 u_\varepsilon(x') \\ \varepsilon\, v(x') \end{pmatrix}$$

are isometries.[2.14] That is: $\nabla^\top \overline{y}_\varepsilon \nabla \overline{y}_\varepsilon = I_2$, where

$$\nabla \overline{y}_\varepsilon = \begin{pmatrix} I_2 \\ 0\ 0 \end{pmatrix} + \varepsilon \begin{pmatrix} 0_2 \\ \nabla^\top v \end{pmatrix} + \varepsilon^2 \begin{pmatrix} \nabla u_\varepsilon \\ 0\ 0 \end{pmatrix} \in \mathbb{R}^{3\times2}.$$

Additionally the following normal vectors are unitary in $\mathbb{R}^3$:

$$b_\varepsilon(x') := \overline{y}_{\varepsilon,1}(x') \wedge \overline{y}_{\varepsilon,2}(x')$$

$$= -\varepsilon \begin{pmatrix} \nabla v \\ 0 \end{pmatrix} + \begin{pmatrix} \varepsilon^3 \nabla u_{\varepsilon2} \cdot (v_{,2}, -v_{,1}) \\ \varepsilon^3 \nabla u_{\varepsilon1} \cdot (-v_{,2}, v_{,1}) \\ 1 + \varepsilon^2 \operatorname{tr} \nabla u_\varepsilon + \varepsilon^4 \det \nabla u_\varepsilon \end{pmatrix}$$

$$= e_3 - \varepsilon \hat{\nabla} v(x') + r_\varepsilon(x'),$$

2.14. The uniform bounds for $\|u_\varepsilon\|_{2,2}$ follow from [FJM06, Theorem 7], equation (181), and those for $\|u_\varepsilon\|_{2,\infty}$ from the explicit construction done in the proof, in particular equations (183), (186) and (190).

where the rest r_ε satisfies

$$\|r_\varepsilon\|_{1,\infty} = \mathcal{O}(\varepsilon^2),$$

by virtue of $\|u_\varepsilon\|_{2,\infty} \leqslant C$ and $\|\nabla v\|_\infty \leqslant C$. Consequently the matrices

$$R_\varepsilon := (\nabla \bar{y}_\varepsilon, b_\varepsilon) = I + \varepsilon \left(\begin{array}{c|c} 0 & -\nabla v \\ \hline \nabla^T v & 0 \end{array} \right) + \underbrace{r_\varepsilon \otimes e_3 + \varepsilon^2 \hat{\nabla} u_\varepsilon}_{=:\tilde{r}_\varepsilon}$$

are in $SO(3)$ for every $x' \in \omega$, with the remaining matrix $\tilde{r}_\varepsilon$ satisfying

$$\|\tilde{r}_\varepsilon\|_{1,\infty} = \mathcal{O}(\varepsilon^2)$$

by the same arguments as before. Now, for some smooth functions a, g_1, $g_2 \in C^\infty(\bar{\omega}; \mathbb{R})$, $g := (g_1, g_2)$ and $d \in L^\infty(\Omega_1; \mathbb{R}^3)$ with $\nabla' d \in L^\infty(\Omega_1; \mathbb{R}^{3\times 2})$ and $D^h \in C^\infty(\bar{\Omega}_1; \mathbb{R}^3)$ to be determined later, set

$$\begin{aligned} y^h(x', x_3) := \ & \bar{y}_\varepsilon(x') + h(x_3 - a(x')) b_\varepsilon(x') + \varepsilon h(g(x'), 0) \\ & + \varepsilon h^2 \int_0^{x_3} d(x', \xi) \, d\xi + D^h(x', x_3). \end{aligned} \tag{2.19}$$

We will prove

$$\mathscr{I}_\alpha^h(y^h) \xrightarrow[h\to 0]{} \mathscr{I}_{lKi}(v).$$

as well as $P_\alpha^h(y^h) \to v$ in $W^{1,2}$ for some constants $R^h \in SO(3)$, $c^h \in \mathbb{R}^3$.

Step 2: *Preliminary computations.*

In order to compute the limit of $\frac{1}{h^{2\alpha-2}} \int_{\Omega_1} W_0(x_3, \nabla_h y^h (I + \varepsilon h B^h))$ we start with the gradient of the recovery sequence:

$$\begin{aligned} \nabla_h y^h = \ & (\nabla \bar{y}_\varepsilon, 0) + h \nabla_h [(x_3 - a) b_\varepsilon] \\ & + \varepsilon h [\hat{\nabla} g + d \otimes e_3] + \nabla_h D^h + o(\varepsilon h). \end{aligned}$$

For the term in h and any $i \in \{1, 2, 3\}$ and $j \in \{1, 2\}$ we have

$$\partial_j [(x_3 - a(x')) b_\varepsilon(x')]_i = \partial_j [(x_3 - a) b_{\varepsilon i}] = (x_3 - a) b_{\varepsilon i, j} - \varepsilon a_{,j} b_{\varepsilon i}.$$

Also: $\frac{1}{h} \partial_3 [(x_3 - a) b_\varepsilon] = \frac{1}{h} b_\varepsilon$, so that

$$\begin{aligned} \nabla_h [(x_3 - a) b_\varepsilon] &= (x_3 - a) \hat{\nabla} b_\varepsilon - b_\varepsilon \otimes \hat{\nabla} a + \frac{1}{h} b_\varepsilon \otimes e_3 \\ &= (a - x_3)(\varepsilon \hat{\nabla}^2 v - \hat{\nabla} r_\varepsilon) - b_\varepsilon \otimes \hat{\nabla} a + \frac{1}{h} b_\varepsilon \otimes e_3. \end{aligned}$$

Substituting back into the gradient yields:

$$\nabla_h y^h = \underbrace{R_\varepsilon + \varepsilon h\left[(a-x_3)\hat{\nabla}^2 v + \hat{\nabla}g + d\otimes e_3 + o(1)\right]}_{=:A^h}$$
$$-hb_\varepsilon \otimes \hat{\nabla}a + \nabla_h D^h. \tag{2.20}$$

Because we intend to use the frame invariance of the energy, we will need the product of $\nabla_h y^h$ with $R_\varepsilon^\top = I + \mathcal{O}(\varepsilon)$. First we have:

$$\varepsilon h R_\varepsilon^\top A^h = \varepsilon h A^h + o(\varepsilon h) = \varepsilon h A^h,$$

where we have subsumed terms $o(\varepsilon h)$ into the $o(1)$ inside A^h. Using $|b_\varepsilon| \equiv 1$ and $\bar{y}_{\varepsilon,i} \perp b_\varepsilon$ we also have $R_\varepsilon^\top b_\varepsilon = e_3$. Therefore

$$R_\varepsilon^\top \nabla_h y^h = I_3 + \underbrace{\varepsilon h A^h - h e_3 \otimes \hat{\nabla}a + R_\varepsilon^\top \nabla_h D^h}_{=:F^h}. \tag{2.21}$$

Step 3: *Convergence of the energies.*

The next step is a Taylor expansion around the identity. Given that the energy is scaled by $(\varepsilon h)^{-2}$, only those terms scaling as εh in (2.21) will remain: anything beyond that will not be seen and anything below will make the energy blow up. This means that we must choose D^h so that $F^h = o(\varepsilon h)$. In [FJM06], [Sch07b] it was possible for the authors to obtain exactly $F^h = 0$ by choosing D^h adequately, but in our case this will not be possible.[2.15] If we set $D^h := h^2 D$ for some smooth D, we have

$$F^h = h[D_{,3} \otimes e_3 + \varepsilon(v_{,1}D_{3,3}, v_{,2}D_{3,3}, -v_{,1}D_{1,3} - v_{,2}D_{2,3}) \otimes e_3$$
$$-e_3 \otimes \hat{\nabla}a + o(\varepsilon)]$$
$$=: h\tilde{F}^h.$$

2.15. Technically, this is due to the fact that the term $h e_3 \otimes \hat{\nabla}a$ is a row in a matrix instead of a column, which makes it impossible to exactly compensate because $R_\varepsilon^\top \nabla_h D^h$ effectively only provides a column vector to work with. Indeed,

$$R_\varepsilon^\top \nabla_h D^h = \nabla_h D^h + \varepsilon \left(\begin{array}{c|c} 0 & \nabla v \\ \hline -\nabla^\top v & 0 \end{array} \right) \nabla_h D^h + \tilde{r}_\varepsilon^\top \nabla_h D^h,$$

so in order to cancel $h e_3 \otimes \hat{\nabla}a$ we must have that the leading term $\nabla_h D^h$ be of order h. But then $\nabla_h D^h = \left(\nabla' D^h, \frac{1}{h}D_{,3}^h\right)$ requires that D^h scale at least as $h^2 \ll \varepsilon h \ll 1$ so we "lose" the first two columns of $\nabla_h D^h$.

This means that we must solve the equations $\tilde{F}^h = o(\varepsilon)$. Although these have no solution the symmetrised version does,[2.16] so that for every smooth choice of a we can pick a bounded D^h such that

$$\tilde{F}^h_s = 0, \text{ and } \tilde{F}^h = \mathcal{O}(1), \tag{2.22}$$

a fact that we will exploit next. By frame invariance, (2.21) and $F^h = h\tilde{F}^h$, we can write

$$
\begin{aligned}
W_0(x_3, & \nabla_h y^h (I + \varepsilon\, h\, B^h)) \\
&= W_0(x_3, R_\varepsilon^\top \nabla_h y^h (I + \varepsilon\, h\, B^h)) \\
&= W_0(x_3, (I + \varepsilon\, h A^h + h\tilde{F}^h)(I + \varepsilon\, h\, B^h)) \\
&= W_0(x_3, I + \underbrace{h(\varepsilon\,(A^h + B^h) + \tilde{F}^h + o(\varepsilon))}_{=:C^h}).
\end{aligned}
$$

Because of (2.22) by our choice of D we need to subtract the antisymmetric part of $\tilde{F}^h$, which we do by means of another rotation (cf. Lemma A.10) and frame invariance:

$$
\begin{aligned}
W_0(x_3, I + h\,C^h) &= W_0\big(x_3, e^{-h\tilde{F}^h_a}(I + h\,C^h)\big) \\
&= W_0(x_3, (I - h\tilde{F}^h_a + \mathcal{O}(h^2))(I + h\,C^h)) \\
&= W_0(x_3, I + h\,C^h - h\tilde{F}^h_a + \mathcal{O}(h^2)) \\
&= W_0(x_3, I + \varepsilon\, h(A^h + B^h) + o(\varepsilon\, h)).
\end{aligned}
$$

Now whenever h is small enough that $I + h\,C^h$ belongs to the neighbourhood of $SO(3)$ where W_0 is twice differentiable, we can apply Taylor's theorem and the fact that Q_3 vanishes on antisymmetric matrices (Lemma A.13) to see that, as $h \to 0$:

$$
\begin{aligned}
\frac{1}{\varepsilon^2 h^2} W_0(x_3, \nabla_h y^h (I + \varepsilon\, h\, B^h)) &= \frac{1}{2} Q_3(x_3, (A^h + B^h)_s) + o(1) \\
&\to \frac{1}{2} Q_3(x_3, A_s + B_s),
\end{aligned}
$$

2.16. Dividing by h we arrive at:

$$
\begin{cases}
D_{1,3} + \varepsilon\, v_{,1} D_{3,3} = a_{,1} + o(\varepsilon), \\
D_{2,3} + \varepsilon\, v_{,2} D_{3,3} = a_{,2} + o(\varepsilon), \\
D_{3,3} - \varepsilon\, v_{,1} D_{1,3} - \varepsilon\, v_{,2} D_{2,3} = o(\varepsilon),
\end{cases}
$$

with solution:

$$D(x', x_3) = x_3 \hat{\nabla} a + x_3 \varepsilon\, \nabla v \cdot \nabla a\, e_3.$$

where
$$A_s = (a - x_3)\,\hat{\nabla}^2 v + \hat{\nabla}_s g + (d \otimes e_3)_s.$$
We choose
$$d(x', x_3) = \mathscr{L}(x_3, (a - x_3)\,\nabla^2 v + \nabla_s g + \check{B}_s) - B_{\cdot 3},$$

with $\mathscr{L}$ the continuous map from Lemma A.12 and $B_{\cdot 3}$ the third column of B. Because the matrix $(a - x_3)\,\nabla^2 v + \nabla_s g + \check{B}_s$ is bounded uniformly in x', by Lemma A.15 the map

$$x \mapsto \int_0^{x_3} \mathscr{L}(\xi, (a - \xi)\,\hat{\nabla}^2 v + \hat{\nabla}_s g + B_s(\xi))\,d\xi$$

is in $W^{1,\infty}(\Omega_1; \mathbb{R}^3)$ and $y^h \in W^{1,2}$ as required (for the derivatives wrt. x' note that v, g are smooth and B independent of x').

Now, all quantities being bounded, by dominated convergence:

$$\mathscr{I}_\alpha^h(y^h) \;\to\; \frac{1}{2}\int_{\Omega_1} Q_3(x_3, (a - x_3)\,\hat{\nabla}^2 v + \hat{\nabla}_s g + (d \otimes e_3)_s + B_s)\,dx$$

$$= \;\frac{1}{2}\int_{\Omega_1} Q_2(x_3, (a - x_3)\,\nabla^2 v + \nabla_s g + \check{B}_s)\,dx.$$

Note that a final step is required to obtain convergence to $\mathscr{I}_{lKi}(v)$.

Step 4: *Convergence of the deformations:* $P_\alpha^h(y^h) \to v$ *in* $W^{1,2}$.
$\underline{}$ We have

$$P_\alpha^h(y^h) = \frac{1}{\varepsilon}\int_{-\frac{1}{2}}^{\frac{1}{2}} y_3^h(x', x_3)\,dx_3,$$

where in (2.19) we defined $y_3^h(x', x_3) = \varepsilon\,v(x') + h\,(x_3 - a(x'))\,b_{\varepsilon 3}(x') + \mathcal{O}(\varepsilon h)$. Then:

$$|P_\alpha^h(y^h) - v|^2 \;=\; \left|\frac{1}{\varepsilon}\int_{-\frac{1}{2}}^{\frac{1}{2}} [\varepsilon v + h(x_3 - a)b_{\varepsilon 3} + \mathcal{O}(\varepsilon h)]\,dx_3 - v\right|^2$$

$$= \;\mathcal{O}(\varepsilon^{-2} h^2),$$

and consequently $\|P_\alpha^h(y^h) - v\|_{0,2} \to 0$. An analogous computation for the derivatives shows strong convergence in $W^{1,2}$.

For the maps $R^h \colon \omega \to SO(3)$, take

$$R^h := R_\varepsilon\, e^{h\bar{F}_a^h}.$$

Then $\|R^h - I\|_{0,2} = \|(I + \mathcal{O}(\varepsilon))(I + \mathcal{O}(h)) - I\|_{0,2} = \mathcal{O}(h^{\alpha-2})$ and:

$$
\begin{aligned}
\|\nabla_h y^h - R^h\|_{0,2} &= \left\| e^{-h\tilde{F}_a^h} R_\varepsilon^\top \nabla_h y^h - I \right\|_{0,2} \\
&= \left\| e^{-h\tilde{F}_a^h} R_\varepsilon^\top \nabla_h y^h - I \right\|_{0,2} \\
&= \|(I - h\tilde{F}_a^h)(I + \varepsilon\, h A^h + h\tilde{F}^h) + o(\varepsilon\, h) - I\|_{0,2} \\
&\leqslant \mathcal{O}(h^{\alpha-1}).
\end{aligned}
$$

Step 5: *Simultaneous convergence.*

Finally, as in [Sch07b, Theorem 3.2], in order for the energy to converge to the true limit, we must pick a and g in (2.19) so as to approximate the minimum $\overline{Q}_2^\star$. This is done with Corollary 2.23, substituting sequences of smooth functions $(a_k)_{k \in \mathbb{N}}$, $(g_k)_{k \in \mathbb{N}}$ for the functions a, g. Then, for each fixed k we have:

$$
\begin{aligned}
\mathscr{J}_\alpha^h(y_k^h) \underset{h \to 0}{\to} \frac{1}{2} \int_{\Omega_1} Q_2(x_3, (a_k - x_3)\nabla^2 v + \nabla_s g_k + \check{B}_s) \\
= \frac{1}{2} \int_\omega \overline{Q}_2^\star(-\nabla^2 v)\, dx' + o(1)_{k \to \infty},
\end{aligned}
$$

and

$$
\|P_\alpha^h(y_k^h) - v\|_{1,2}^2 \leqslant C(k)\,\varepsilon^{-2} h^2.
$$

And by a diagonal argument we can find $(y^h)_{h>0}$ whose energy converges to $\mathscr{J}_{lKi}(v)$ while maintaining the convergence of the deformations. $\quad\square$

Theorem 2.10. (Upper bound, von Kármán regime) *Let $\alpha = 3$ and consider displacements $(u, v) \in X_{\alpha=3} := W^{1,2}(\omega; \mathbb{R}^2) \times W^{1,2}(\omega; \mathbb{R})$. There exists a sequence $(y^h)_{h>0} \subset Y$ which P_θ^h-converges to (u, v) such that*

$$
\lim_{h \to 0} \frac{1}{\theta} \mathscr{J}_\alpha^h(y^h) = \mathscr{J}_{vK}^\theta(u, v),
$$

with $\mathscr{J}_{vK}^\theta$ defined as in (2.10) as

$$
\mathscr{J}_{vK}^\theta(u, v) := \frac{1}{2} \int_\omega \overline{Q}_2\left(\theta^{1/2}\left(\nabla_s u + \frac{1}{2}\nabla v \otimes \nabla v\right), -\nabla^2 v\right)
$$

over $X_{\alpha=3}^0 = W^{1,2}(\omega; \mathbb{R}^2) \times W^{2,2}(\omega; \mathbb{R})$ and as ∞ elsewhere.

Proof. In order to build the recovery sequence $(y^h)_{h>0}$ we will use the map $\mathscr{L}: \left(-\tfrac{1}{2}, \tfrac{1}{2}\right) \times \mathbb{R}^{2\times2} \to \mathbb{R}^3$ given by Lemma A.12, which for each t realises the minimum of $Q_3(t, \hat{A} + c \otimes e_3), A \in \mathbb{R}^{2\times2}$, i.e.

$$Q_2(t,A) = Q_3(t, \hat{A} + \mathscr{L}(t,A) \otimes e_3) = Q_3(t, \hat{A} + (\mathscr{L}(t,A) \otimes e_3)_s),$$

where the last equality follows from the fact that Q_2 vanishes on antisymmetric matrices (Lemma A.13). Recall that $\mathscr{L}(t,\cdot)$ is linear for every t and that $|\mathscr{L}(t,A)| \lesssim |A|$ uniformly in t (Lemma A.15).

The functional $\mathscr{I}_{vK}^\theta$ is clearly continuous in $X_\alpha^0 = W^{1,2}(\omega; \mathbb{R}^2) \times W^{2,2}(\omega; \mathbb{R})$ with the *strong* topologies, so we may apply Lemma A.27 and it is enough to consider $(u, v) \in C^\infty(\overline{\omega}; \mathbb{R}^2) \times C^\infty(\overline{\omega}; \mathbb{R})$, which is dense in X_α^0. We define:

$$
y^h(x',x_3) := \begin{pmatrix} x' \\ hx_3 \end{pmatrix} + \begin{pmatrix} \theta h^2 u(x') \\ \sqrt{\theta}\, h\, v(x') \end{pmatrix} - \sqrt{\theta}\, h^2 x_3 \begin{pmatrix} \nabla v(x') \\ 0 \end{pmatrix}
$$
$$
+ \theta h^3 d(x',x_3) \tag{2.23}
$$

where $d \in W^{1,\infty}(\Omega_1; \mathbb{R}^3)$ is a vector field to be determined along the proof.

Step 1: *Approximation of the energy.*

A direct computation yields

$$
\begin{aligned}
\nabla_h y^h &= I + \left(\begin{array}{c|c} \theta h^2 \nabla u & -h\sqrt{\theta}\,\nabla v \\ \hline h\sqrt{\theta}\,\nabla^\top v & 0 \end{array} \right) - h^2 \theta \left(\begin{array}{c|c} x_3 \theta^{-\frac{1}{2}} \nabla^2 v & 0 \\ \hline 0 & 0 \end{array} \right) \\
&\quad + h^2 \theta\, \partial_3 d \otimes e_3 + \mathcal{O}(h^3) \\
&= I + h\sqrt{\theta}\, \underbrace{(e_3 \otimes \hat{\nabla} v - \hat{\nabla} v \otimes e_3)}_{E} \\
&\quad + h^2 \theta\, \underbrace{\left(\hat{\nabla} u - x_3 \theta^{-\frac{1}{2}} \hat{\nabla}^2 v + \partial_3 d \otimes e_3 \right)}_{F} + \mathcal{O}(h^3).
\end{aligned}
$$

For later use we note here the product:

$$
\begin{aligned}
\nabla_h^\top y^h \nabla_h y^h &= \left(I + h\sqrt{\theta}\, E^\top + h^2 \theta F^\top \right)\left(I + h\sqrt{\theta}\, E + h^2 \theta F \right) + \mathcal{O}(h^3) \\
&= I + h\sqrt{\theta}\, \underbrace{2 E_s}_{=0} + h^2 \theta\, \underbrace{(2F_s + E^\top E)}_{N} + \mathcal{O}(h^3),
\end{aligned}
$$

where we used that E is antisymmetric. For any matrix M with positive determinant we have the polar decomposition $M = U\sqrt{M^\top M} = U\sqrt{I+P}$, with $U\in SO(3)$ and $P=M^\top M - I$. By the frame invariance of the energy and a Taylor expansion around the identity of the square root

$$
\begin{aligned}
W_0(x_3, M) &= W_0(x_3, \sqrt{M^\top M})\\
&= W_0\big(x_3, I + \tfrac{1}{2}(M^\top M - I) + o(|M^\top M - I|)\big),
\end{aligned}
$$

and, assuming that a Taylor expansion of W_0 around the identity can be carried, i.e. that M is close enough to $SO(3)$, this is equal to:

$$
\frac{1}{2} Q_3\big(x_3, \tfrac{1}{2}(M^\top M - I)\big) + o(|M^\top M - I|^2).
$$

In view of the definition of W_0, we set

$$
M^h := \nabla_h y^h (I + h^2 \tilde{B}^h),
$$

where $\tilde{B}^h = \sqrt{\theta}\, B^h \to \tilde{B} = \sqrt{\theta}\, B$ in L^∞. Then

$$
\begin{aligned}
(M^h)^\top M^h &:= [\nabla_h y^h (I + h^2 \tilde{B}^h)]^\top [\nabla_h y^h (I + h^2 \tilde{B}^h)]\\
&= (I + h^2 (\tilde{B}^h)^\top) \nabla_h^\top y^h \nabla_h y^h (I + h^2 \tilde{B}^h).\\
&= (I + h^2 (\tilde{B}^h)^\top)(I + h^2 \theta N)(I + h^2 \tilde{B}^h) + \mathcal{O}(h^3)\\
&= I + h^2 \theta N + h^2 2 \tilde{B}^h_s + \mathcal{O}(h^3)\\
&= I + h^2 \theta N + h^2 2 \tilde{B}_s + o(h^2).
\end{aligned}
$$

To compute the first term in h^2, $N = 2 F_s + E^\top E$, we have

$$
2 F_s = 2\left(\hat{\nabla}_s u - x_3 \theta^{-1/2} \hat{\nabla}^2 v + (\partial_3 d \otimes e_3)_s\right),
$$

and:[2.17]

$$
\begin{aligned}
E^\top E &= (\hat{\nabla} v \otimes e_3 - e_3 \otimes \hat{\nabla} v)(e_3 \otimes \hat{\nabla} v - \hat{\nabla} v \otimes e_3)\\
&= \hat{\nabla} v \otimes \hat{\nabla} v + |\hat{\nabla} v|^2 e_3 \otimes e_3.
\end{aligned}
$$

2.17. We use the identities $(c \otimes e_3)(e_3 \otimes c) = c \otimes c$, $(c \otimes e_3)(c \otimes e_3) = c_3\, c \otimes e_3$, $(e_3 \otimes c)(e_3 \otimes c) = c_3 e_3 \otimes c$ and $(e_3 \otimes c)(c \otimes e_3) = |c|^2 e_3 \otimes e_3$.

Since these quantities are independent of h, for sufficiently small h the product $(M^h)^\top M^h$ does lie close enough to $SO(3)$ and we can perform the desired Taylor expansion:

$$
\begin{aligned}
W^h(x_3, \nabla_h y^h) &= W_0(x_3, \nabla_h y^h (I + h^2 \tilde{B}^h)) \\
&= W_0\left(x_3, ((M^h)^\top M^h)^{1/2}\right) \\
&= \frac{1}{2} Q_3\left(x_3, \frac{1}{2}[(M^h)^\top M^h - I]\right) + o(|(M^h)^\top M^h - I|^2).
\end{aligned}
$$

Define now $\hat{G}_0 := \theta\left(\hat{\nabla}_s u + \frac{1}{2}\hat{\nabla} v \otimes \hat{\nabla} v\right)$, $\hat{G}_1 := -\theta^{1/2} \hat{\nabla}^2 v$ as in Lemma A.31. Bringing the previous computations together we obtain:

$$
\begin{aligned}
\frac{1}{2}[(M^h)^\top M^h - I] &= h^2 \Big[\hat{G}_0 - x_3 \hat{G}_1 + \overset{\hat{2}}{\check{B}}_s \\
&\quad + \underbrace{\sqrt{\theta}(B(t)._3 \otimes e_3)_s + \frac{\theta}{2}|\hat{\nabla} v|^2 e_3 \otimes e_3 + \theta(\partial_3 d \otimes e_3)_s}_{H} \Big] \\
&\quad + o(h^2),
\end{aligned}
$$

hence

$$
\begin{aligned}
&\frac{1}{h^4}\left[Q_3\left(x_3, \frac{1}{2}((M^h)^\top M^h - I)\right) + o(|(M^h)^\top M^h - I|^2) \right] \\
&\quad = Q_3\left(x_3, \hat{G}_0 - x_3 \hat{G}_1 + \sqrt{\theta}\,\overset{\hat{}}{\check{B}}_s + H\right) + o(1),
\end{aligned}
$$

We now choose the vector field d to cancel one term and attain the minimum for the others by solving for $\partial_3 d$ in:

$$
H \overset{!}{=} \left(\mathcal{L}\left(x_3, G_0 - x_3 G_1 + \sqrt{\theta}\,\check{B}_s(x_3)\right) \otimes e_3\right)_s,
$$

that is:

$$
\theta^{-1/2} B(t)._3 + \frac{1}{2}|\hat{\nabla} v|^2 e_3 + \partial_3 d(x', x_3) = \frac{1}{\theta}\mathcal{L}\left(x_3, G_0 - x_3 G_1 + \sqrt{\theta}\,\check{B}_s(x_3)\right).
$$

Consequently, we set:

$$
\begin{aligned}
d(x', x_3) :=\ & -\frac{1}{2}|\hat{\nabla} v|^2 x_3 e_3 \\
& + \frac{1}{\theta}\int_0^{x_3} \mathcal{L}\left(t, G_0 - t G_1 + \sqrt{\theta}\,\check{B}_s(t)\right) - \sqrt{\theta}\, B(t)._3 \, dt,
\end{aligned}
$$

and we obtain

$$Q_3\big(x_3,\hat{G}_0-x_3\hat{G}_1+\sqrt{\theta}\,B_s(x_3)+H\big)=Q_2\big(x_3,G_0-x_3G_1+\sqrt{\theta}\,\check{B}_s(x_3)\big).$$

In order to check that $d\in W^{1,\infty}(\Omega_1;\mathbb{R}^3)$ we note first that $G_0(x'),G_1(x')$ are uniformly bounded over ω thanks to u_i,v being $C^\infty(\overline{\omega})$ and $B_s\in L^\infty$. Now, by Lemma A.15 the map $\mathscr{L}$ is bounded uniformly in t over sets of symmetric bounded matrices, so that

$$\|d\|_{0,\infty}\leqslant C\,\|\nabla v\|_{0,\infty}+\theta^{-1}C(\mathscr{L},u,v,B)\leqslant C.$$

For the derivative $\partial_i d,\ i\in\{1,2\}$ we obtain

$$\begin{aligned}\|\partial_i d\|_{0,\infty}\ \lesssim\ & \|\nabla v\|_{0,\infty}\|\nabla^2 v\|_{0,\infty}\\ &+\theta^{-1}C(\mathscr{L},u,v,B)\,(\|\partial_i G_0\|_{0,\infty}+\|\partial_i G_1\|_{0,\infty})\end{aligned}$$

and for the third derivative

$$\|\partial_3 d\|_{0,\infty}\leqslant\theta^{-1}C(\mathscr{L},u,v,B).$$

Step 2: *Convergence.*

By the previous step we have $\frac{1}{\theta h^4}\,W_0(x_3,\nabla_h\,y^h)\to\frac{1}{2\theta}Q_2\big(x_3,G_0-x_3G_1+\sqrt{\theta}\,\check{B}_s\big)$ a.e. as $h\to 0$, and the sequence is uniformly bounded so we can integrate over the domain and pass to the limit:

$$\begin{aligned}\frac{1}{\theta h^4}\int_{\Omega_1}W^h(x_3,\nabla_h y^h)\ &\to\ \frac{1}{2\theta}\int_{\Omega_1}Q_2\big(x_3,G_0-x_3G_1+\sqrt{\theta}\,\check{B}_s\big)\\ &=\ \frac{1}{2}\int_\omega\overline{Q}_2(\theta^{1/2}(\nabla_s u+\tfrac{1}{2}\nabla v\otimes\nabla v),-\nabla^2 v).\end{aligned}$$

Step 3: *Convergence of the recovery sequence: $P_\theta^h(y)\to(u,v)$ in X_α as* $h\to 0$.

For the in-plane displacements we have by definition (2.8) of P_θ^h and (2.23):

$$\begin{aligned}u^h(x')\ &=\ \frac{1}{\theta h^2}\int_{-1/2}^{1/2}(y^h(x',x_3)-x')\,dx_3\\ &=\ u(x')+h\int_{-1/2}^{1/2}d'(x',x_3)\,dx_3,\end{aligned}$$

where the $W^{1,2}$-norm of the last term goes to zero with h because $d \in W^{1,\infty}(\Omega_1)$. Again by (2.8) and (2.23) we have for the out-of-plane displacements

$$v^h(x') = \frac{1}{\sqrt{\theta}\,h} \int_{-1/2}^{1/2} y_3^h(x',x_3)\,dx_3$$

$$= v(x') + h^2 \sqrt{\theta} \int_{-1/2}^{1/2} d(x',x_3)\,dx_3.$$

As before, the second term goes to zero in $W^{1,2}$ and consequently $(u^h, v^h) \to (u, v)$ in $W^{1,2}$.

For the rotations take

$$R^h := e^{\sqrt{\theta}\,h E_a}.$$

Then

$$\begin{aligned}
\|\nabla_h y^h - R^h\|_{0,2} &= \left\| e^{-\sqrt{\theta}\,h E_a} \nabla_h y^h - I \right\|_{0,2} \\
&= \left\| \left(I - \sqrt{\theta}\,h E_a\right)\left(I + \sqrt{\theta}\,h E_a\right) + \mathcal{O}(h^2) - I \right\|_{0,2} \\
&= \mathcal{O}(h^2),
\end{aligned}$$

and $\|R^h - I\| = \mathcal{O}(h)$ as required. $\qquad\square$

In the next result, there is a departure from the analogous functional in [FJM06] beyond the dependence on the out-of-plane component x_3. In the preceding cases, if one sets $Q_2(t, A) \equiv Q_2(A)$, and $B \equiv 0$ then the same functionals are obtained as in that work. However, in the regime $\alpha > 3$ their limit has no membrane term, but we have $\overline{Q}_2(\nabla_s u, -\nabla^2 v) = \frac{1}{2}\int Q_2(\nabla_s u) + \frac{1}{24}\int Q_2(\nabla^2 v)$, *with* the membrane term. The reason is that [FJM06] discard the in-plane displacements by minimising them away. In their proofs, they drop the first term in the lower bound and build the recovery sequence with no u term in $h^{\alpha-1}$.

Note that it is by keeping the membrane term that our model is able to take into account and respond to the pre-stressing (internal misfit) B^h, e.g. compressive or tensile stresses in wafers.

Theorem 2.11. (Upper bound, linearised von Kármán regime) *Let $\alpha > 3$ and consider displacements $(u, v) \in X_\alpha := W^{1,2}(\omega; \mathbb{R}^2) \times W^{1,2}(\omega; \mathbb{R})$. There exists a sequence $(y^h)_{h>0} \subset Y$ which P_α^h-converges to (u, v) such that*

$$\lim_{h \to 0} \mathscr{I}_\alpha^h(y^h) = \mathscr{I}_{lvK}(u, v),$$

with $\mathscr{I}_{lvK}$ defined as in Theorem 2.6 by

$$\mathscr{I}_{lvK}(u, v) := \frac{1}{2} \int_\omega \overline{Q}_2(\nabla_s u, -\nabla^2 v) \, dx'$$

on X_α^0 and by $+\infty$ elsewhere.

Proof. We follow closely the notation and path of proof of Theorem 2.10. By a standard density argument it is enough to consider $(u, v) \in X_\alpha \cap C^\infty(\overline{\omega})$. Define

$$y^h(x', x_3) := \begin{pmatrix} x' \\ h x_3 \end{pmatrix} + \begin{pmatrix} h^{\alpha-1} u(x') \\ h^{\alpha-2} v(x') \end{pmatrix} - h^{\alpha-1} x_3 \begin{pmatrix} \nabla v(x') \\ 0 \end{pmatrix} + h^\alpha d(x', x_3),$$

with $d \in W^{1,\infty}(\Omega_1; \mathbb{R}^3)$. Then

$$\nabla_h y^h = I + h^{\alpha-2} \underbrace{(e_3 \otimes \hat{\nabla} v - \hat{\nabla} v \otimes e_3)}_{=:E} + h^{\alpha-1} \underbrace{(\hat{\nabla} u - x_3 \hat{\nabla}^2 v + \partial_3 d \otimes e_3)}_{=:F}$$
$$+ \mathcal{O}(h^\alpha),$$

and, using that $E_s = 0$:

$$\nabla_h^\top y^h \nabla_h y^h = (I + h^{\alpha-2} E^\top + h^{\alpha-1} F^\top)(I + h^{\alpha-2} E + h^{\alpha-1} F) + \mathcal{O}(h^\alpha)$$
$$= I + 2 h^{\alpha-1} F_s + o(h^{\alpha-1}).$$

Define now $M^h := \nabla_h y^h (I + h^{\alpha-1} B^h)$. A few computations lead to

$$\frac{1}{2}[(M^h)^\top M^h - I] = h^{\alpha-1}(F_s + B_s) + o(h^{\alpha-1}),$$

from which follows, after a Taylor approximation (recall from the proof of Theorem 2.10, that this can be done for sufficiently small h):

$$\frac{1}{h^{2\alpha-2}} W^h(x_3, \nabla_h y^h) = \frac{1}{2 h^{2\alpha-2}} [Q_3(x_3, [(M^h)^\top M^h - I]/2)$$
$$+ o(|(M^h)^\top M^h - I|^2)]$$
$$= \frac{1}{2} Q_3(x_3, F_s + B_s) + o(1).$$

Picking d such that:

$$((B(x_3)._3 + \partial_3 d) \otimes e_3)_s = (\mathscr{L}(x_3, \nabla u - x_3 \nabla^2 v + \check{B}_s(x_3)) \otimes e_3)_s,$$

e.g.

$$d(x', x_3) := \int_0^{x_3} \mathscr{L}(t, \nabla_s u - t \nabla^2 v + \check{B}_s(t)) - B(t)._3 \, dt,$$

the term with $\mathscr{L}$ in Q_2 cancels out and we obtain

$$Q_3(x_3, F_s + B_s) = Q_2(x_3, \nabla_s u - x_3 \nabla^2 v + \check{B}_s(x_3)).$$

Note that as proved in Theorem 2.10, the properties of $\mathscr{L}$ imply that the function $d \in W^{1,\infty}(\Omega_1; \mathbb{R}^3)$ so the previous computations are justified. We have therefore

$$\frac{1}{h^{2\alpha-2}} W_0(x_3, \nabla_h y^h) \to \frac{1}{2} Q_2(x_3, \nabla_s u - x_3 \nabla^2 v + \check{B}_s(x_3)) \text{ a.e. in } \omega,$$

and also $Q_2(x_3, A) \lesssim |A|^2$ because Q_3 is in L^∞ (Assumption 2.2.b). Because $u_i, v \in C^\infty(\overline{\omega})$ and $B_s \in L^\infty$, all arguments of Q_2 are uniformly bounded and we can apply dominated convergence to conclude:

$$\frac{1}{h^{2\alpha-2}} \int_{\Omega_1} W_0(x_3, \nabla_h y^h) \xrightarrow[h \downarrow 0]{} \frac{1}{2} \int_{\Omega_1} Q_2(x_3, \nabla_s u - x_3 \nabla^2 v + \check{B}_s(x_3)) \, dx$$

$$= \frac{1}{2} \int_\omega \overline{Q}_2(\nabla_s u, -\nabla^2 v) \, dx'.$$

It remains to prove that indeed $P_\alpha^h(y^h) \to (u, v)$ in X_α. We begin with the out-of-plane displacements

$$v^h(x') := \frac{1}{h^{\alpha-2}} \int_{-1/2}^{1/2} y_3^h(x', x_3) \, dx_3 = v(x') + h^2 \int_{-1/2}^{1/2} d(x', x_3) \, dx_3,$$

which converges to v in $W^{1,2}$ because $d \in W^{1,\infty}$. Analogously:

$$u^h(x') := \frac{1}{h^{\alpha-1}} \int_{-1/2}^{1/2} (y^{h'}(x', x_3) - x') \, dx_3 = u(x') + h^2 \int_{-1/2}^{1/2} d(x', x_3) \, dx_3$$

converges to u in $W^{1,2}$. Finally, with the choice $R^h := e^{h^{\alpha-2}E_a}$ we have:

$$
\begin{aligned}
\|\nabla_h y^h - R^h\|_{0,2} &= \|e^{-h^{\alpha-2}E_a}\nabla_h y^h - I\|_{0,2} \\
&= \|(I - h^{\alpha-2}E_a)(I + h^{\alpha-2}E_a) + \mathcal{O}(h^{\alpha-1}) - I\|_{0,2} \\
&= \mathcal{O}(h^{\alpha-1}).
\end{aligned}
$$

$\square$

Together with the compactness results of Appendix A.6, this completes the arguments necessary to show convergence of the functionals, of their (almost) minimisers and of their minima. Next, we investigate the transition between $\alpha < 3$ and $\alpha > 3$.

2.4 Γ-convergence of the interpolating theory

Notation. *Throughout this section we write $A_\theta := \nabla_s u_\theta + \frac{1}{2}\nabla v_\theta \otimes \nabla v_\theta$ for the strain induced by a pair of displacements (u_θ, v_θ). As before, $\theta > 0$.*

We now set to prove Theorem 2.7, which states that the functional of generalised von Kármán type that we found in the preceding section,

$$
\mathscr{J}^\theta_{vK}(u_\theta, v_\theta) := \frac{1}{2}\int_\omega \int_{-1/2}^{1/2} Q_2\big(x_3, \sqrt{\theta}\, A_\theta - x_3 \nabla^2 v_\theta + \check{B}(x_3)\big)\, dx_3\, dx',
$$

interpolates between the two adjacent regimes as $\theta \to \infty$ or $\theta \to 0$. As θ approaches infinity, we expect the optimal energy configurations to approach those of the linearised Kirchhoff model, whereas with θ tending to zero they should approach the linearised von Kármán model.

For this section we must restrict ourselves to spaces where Korn-Poincaré type inequalities hold.

Definition 2.12. *Let*

$$
X_u := \left\{ u \in W^{1,2}(\omega; \mathbb{R}^2) : \int_\omega \nabla_a u = 0 \text{ and } \int_\omega u = 0 \right\},
$$

and

$$
X_v := \left\{ v \in W^{2,2}(\omega; \mathbb{R}) : \int_\omega \nabla v = 0 \text{ and } \int_\omega v = 0 \right\}.
$$

We set $X_w := X_u \times X_v$ with the weak topologies.

Additionally, from now on we assume that ω **is symmetric** (i.e. $x \in \omega \Leftrightarrow -x \in \omega$).[2.18] This condition is inessential but simplifies some of the computations by removing the need to shift functions. Finally, for the limit $\theta \to \infty$ we require that ω **be convex** and recall the definition of the space of maps with singular Hessian

$$W_{sh}^{2,2}(\omega) := \{v \in W^{2,2}(\omega; \mathbb{R}) : \det \nabla^2 v = 0 \text{ a.e.}\}.$$

Remark 2.13. Working in $X_u \times X_v$ does not affect the energies: First we can always add an infinitesimal rigid motion to u and and any affine function to v without changing $\nabla_s u$ or $\nabla^2 v$. Second, although the nonlinear term $\nabla v \otimes \nabla v$ *does* change after adding an affine function, the extra terms appearing happen to be a symmetric gradient which can be absorbed into $\nabla_s u$ with a little help: For any $g(x) = a \cdot x + b$ with $a, b \in \mathbb{R}^2$, we have

$$\begin{aligned} \nabla(v+g) \otimes \nabla(v+g) &= \nabla v \otimes \nabla v + a \otimes a + a \otimes \nabla v + \nabla v \otimes a \\ &= \nabla v \otimes \nabla v + \nabla_s z \end{aligned} \tag{2.24}$$

where we set $z(x) := (2\,v(x) + a \cdot x)\,a \in W^{2,2}(\omega; \mathbb{R}^2)$. Therefore, for any fixed $u \in W^{1,2}(\omega; \mathbb{R}^2)$, $v \in W^{2,2}(\omega)$ one can choose $g(x) = -[(\nabla v)_\omega \cdot x + (v)_\omega]$ and define

$$\tilde{u} = u + z + r, \quad \tilde{v} = v + g,$$

with $r(x) = R\,x + c$, for constants $R := \frac{-1}{|\omega|} \int_\omega \nabla_a u + \nabla_a z \, dx \in \mathbb{R}_{ant}^{2 \times 2}$ and $c := \frac{-1}{|\omega|} \int_\omega u(x) + z(x) + R\,x \, dx$. For $\tilde{u}, \tilde{v}$ we then have on the one hand $\int \tilde{u} = 0, \int \nabla_a \tilde{u} = 0$ and $\int \tilde{v} = 0, \int \nabla \tilde{v} = 0$ and on the other (note that $\nabla_s r = 0$):

$$\mathcal{J}_{vK}^\theta(u,v) = \mathcal{J}_{vK}^\theta(\tilde{u} - z - r, \tilde{v} - g) \overset{(2.24)}{=} \mathcal{J}_{vK}^\theta(\tilde{u} - r, \tilde{v}) = \mathcal{J}_{vK}^\theta(\tilde{u}, \tilde{v})$$

as desired.

Our first theorem identifies the types of convergence required in order to obtain precompactness of sequences of bounded energy. We use these definitions of convergence for the computation of the Γ-limits.

2.18. Actually, we only need the barycenter to be the origin so that $\int_\omega x' \, dx' = 0$, but for consistency with Section 3.2.1 we require symmetry.

Theorem 2.14. (Compactness) *Let $(u_\theta, v_\theta)_{\theta > 0}$ be a sequence in X_w with finite energy*

$$\sup_{\theta > 0} \mathscr{I}_{vK}^\theta(u_\theta, v_\theta) \leqslant C.$$

Then:

1. *The sequence $(v_\theta)_{\theta \uparrow \infty}$ is weakly precompact in $W^{2,2}(\omega)$ and the weak limit is in $X_v \cap W_{sh}^{2,2}(\omega)$. Additionally $(u_\theta)_{\theta \uparrow \infty}$ is weakly precompact in $W^{1,2}(\omega; \mathbb{R}^2)$.*
2. *The sequence $(\theta^{1/2} u_\theta, v_\theta)_{\theta \downarrow 0}$ is weakly precompact in $W^{1,2}(\omega; \mathbb{R}^2) \times W^{2,2}(\omega)$ and the weak limit is in $X_u \times X_v$.*

Proof. By assumption:

$$C \geqslant \int_\omega \int_{-1/2}^{1/2} Q_2\big(x_3, \sqrt{\theta} A_\theta - x_3 \nabla^2 v_\theta + \check{B}(x_3)\big) \, dx_3 \, dx'$$

and by Lemma A.14 we have the uniform lower bound

$$Q_2(x_3, F) \gtrsim |F|^2 \text{ for all symmetric } F \text{ and } x_3 \in \left(-1/2, 1/2\right),$$

so that $\int_{-1/2}^{1/2} Q_2(x_3, F(x_3)) \gtrsim \int_{-1/2}^{1/2} |F(x_3)|^2$. Now split the inner integral in half, and normalize to use Jensen's inequality. In the upper half:

$$C \geqslant \int_\omega 2 \int_0^{1/2} Q_2\big(x_3, \sqrt{\theta} A_\theta - x_3 \nabla^2 v_\theta + \check{B}_s(x_3)\big) \, dx_3 \, dx'$$

$$\gtrsim \int_\omega 2 \int_0^{1/2} \big|\sqrt{\theta} A_\theta - x_3 \nabla^2 v_\theta + \check{B}_s(x_3)\big|^2 \, dx_3 \, dx'$$

$$\gtrsim \int_\omega \left|2 \int_0^{1/2} \sqrt{\theta} A_\theta - x_3 \nabla^2 v_\theta + \check{B}_s(x_3) \, dx_3\right|^2 \, dx'$$

$$= \int_\omega \left|\sqrt{\theta} A_\theta - \frac{1}{4} \nabla^2 v_\theta + c\right|^2 \, dx'$$

$$\gtrsim \left\|\sqrt{\theta} A_\theta - \frac{1}{4} \nabla^2 v_\theta\right\|_{0,2,\omega}^2 - c^2 |\omega|.$$

An analogous computation for the lower half of the interval results in

$$C \geqslant \left\|\sqrt{\theta} A_\theta + \frac{1}{4} \nabla^2 v_\theta\right\|_{0,2},$$

and bringing both bounds together we obtain:

$$\left\|\sqrt{\theta}\,A_\theta\right\|_{0,2}\leqslant C \quad\text{and}\quad \|\nabla^2 v_\theta\|_{0,2}\leqslant C. \tag{2.25}$$

Two applications of Poincaré's inequality to the second bound yield:

$$\|v_\theta\|_{2,2}\leqslant C \text{ for all } \theta>0.$$

Therefore a subsequence (not relabeled) $v_\theta \rightharpoonup v$ for some $v \in W^{2,2}(\omega)$. Since X_v is convex, it is weakly closed and $v \in X_v$. Now consider (2.25) again and observe that with the Sobolev embedding $W^{1,2}(\omega)\hookrightarrow L^4(\omega)$ we know that

$$\|\nabla v_\theta \otimes \nabla v_\theta\|_{0,2}=\|\nabla v_\theta\|_{0,4}^2\lesssim\|\nabla v_\theta\|_{1,2}^2\leqslant\|v_\theta\|_{2,2}^2\leqslant C.$$

An application of the triangle inequality implies then

$$\left\|\sqrt{\theta}\,\nabla_s u_\theta\right\|_{0,2}\leqslant C+C\sqrt{\theta}, \tag{†}$$

so, by Korn-Poincaré (Corollary A.22), the sequence $(u_\theta)_{\theta>0}$ is bounded in $W^{1,2}$ when $\theta\to\infty$ and there exists a subsequence (not relabeled) $u_\theta \rightharpoonup u$ for some $u\in X_u$ (again by convexity of X_u).

Now if $z_\varepsilon \rightharpoonup z$ in $W^{1,2}(\omega;\mathbb{R}^2)$, by the compact Sobolev embedding $W^{1,2}\hookrightarrow L^4$ we have $z_\varepsilon \to z$ in L^4 and

$$
\begin{aligned}
\int_\omega |z_\varepsilon \otimes z_\varepsilon - z\otimes z|^2\,\mathrm{d}x &\leqslant \int_\omega |z_\varepsilon \otimes (z_\varepsilon - z)|^2 + |(z_\varepsilon - z)\otimes z|^2\,\mathrm{d}x\\
&\leqslant \int_\omega |z_\varepsilon|^2 |z_\varepsilon - z|^2 + |z_\varepsilon - z|^2 |z|^2\,\mathrm{d}x\\
&\leqslant \|z_\varepsilon\|_{0,4}^2\,\|z_\varepsilon - z\|_{0,4}^2\\
&\quad + \|z_\varepsilon - z\|_{0,4}^2\,\|z\|_{0,4}^2 \xrightarrow[\varepsilon\to 0]{} 0.
\end{aligned}
\tag{2.26}
$$

So $\nabla v_\theta \otimes \nabla v_\theta \to \nabla v\otimes \nabla v$ in L^2 and from (2.25) and lower semicontinuity of the norm we deduce

$$\left\|\nabla_s u+\tfrac{1}{2}\nabla v\otimes \nabla v\right\|_{0,2}\leqslant \liminf_{\theta\to\infty}\|A_\theta\|_{0,2}=0.$$

By [FJM06, Proposition 9] $v\in W^{2,2}_{sh}(\omega)$ and this concludes the proof of the first statement.

For the second statement we take $\theta \downarrow 0$. It only remains to prove pre-compactness for u_θ since the previous computation for $(v_\theta)_{\theta>0}$ applies for all θ. But it follows directly from (†) above: again with Corollary A.22, the sequence $(\theta^{1/2} u_\theta)_{\theta>0}$ is bounded in $W^{1,2}$, so it contains a weakly convergent subsequence $\theta^{1/2} u_\theta \rightharpoonup u$ and the limit is in the convex set X_u. □

We begin the proof of Γ-convergence in Theorem 2.7 with the lower and upper bound for the passage from $\alpha = 3$ to $\alpha < 3$.

Theorem 2.15. (Lower bound, von Kármán to linearised Kirchhoff)
Assume ω is convex and let $(u_\theta, v_\theta)_{\theta>0}$ be a sequence in X_w such that $v_\theta \rightharpoonup v$ in X_v as $\theta \to \infty$. Then

$$\liminf_{\theta \uparrow \infty} \mathscr{I}_{vK}^\theta(u_\theta, v_\theta) \geqslant \mathscr{I}_{lKi}(v).$$

Proof. We must consider two cases.

Case 1: $v \in X_v^0 := X_v \cap W_{sh}^{2,2}(\omega)$, hence $\mathscr{I}_{lKi}(v) < \infty$.

We can minimise the inner integral pointwise and obtain a lower bound:

$$\mathscr{I}_{vK}^\theta(u_\theta, v_\theta) = \frac{1}{2} \int_\omega \int_{-1/2}^{1/2} Q_2\left(x_3, \sqrt{\theta}\, A_\theta - x_3 \nabla^2 v_\theta + \check{B}(x_3)\right) dx_3 \, dx'$$

$$\geqslant \frac{1}{2} \int_\omega \min_{A \in \mathbb{R}^{2\times 2}} \int_{-1/2}^{1/2} Q_2(x_3, A - x_3 \nabla^2 v_\theta + \check{B}(x_3)) \, dx_3 \, dx'$$

$$= \mathscr{I}_{lKi}(v_\theta).$$

By Remark 2.3 $\overline{Q}_2^\star$ is convex, therefore $\mathscr{I}_{lKi}$ is w.s.l.s.c. in L^2 [FL07, Theorem 5.14], and we have by the convergence $v_\theta \rightharpoonup v$ in $W^{2,2}$:

$$\liminf_{\theta \uparrow \infty} \mathscr{I}_{vK}^\theta(u_\theta, v_\theta) \geqslant \liminf_{\theta \uparrow \infty} \mathscr{I}_{lKi}(v_\theta) \geqslant \mathscr{I}_{lKi}(v).$$

Case 2: $v \notin X_v^0$, hence $\mathscr{I}_{lKi}(v) = +\infty$.

We need to show that $\mathscr{I}_{vK}^\theta(u_\theta, v_\theta)$ diverges. To prove this by contradiction assume that

$$\liminf_{\theta \uparrow \infty} \mathscr{I}_{vK}^\theta(u_\theta, v_\theta) < \infty.$$

But then there exists some subsequence (not relabelled) such that

$$\sup_{\theta>0} \mathscr{I}_{vK}^\theta(u_\theta, v_\theta) \leqslant C,$$

and by Theorem 2.14 $v \in X_v^0$ in contradiction to the assumption.   □

Theorem 2.16. (Upper bound, von Kármán to linearised Kirchhoff)
Assume ω is convex. Set $X_v^0 := X_v \cap W_{sh}^{2,2}(\omega)$ and fix some displacement $v \in X_v$. There exists a sequence $(u_\theta, v_\theta)_{\theta\uparrow\infty} \subset X_w$ such that $v_\theta \rightharpoonup v$ in $W^{2,2}(\omega)$ and $\limsup_{\theta\uparrow\infty} \mathcal{I}_{vK}^\theta(u_\theta, v_\theta) \leqslant \mathcal{I}_{lKi}(v)$.

Proof. By Theorem 2.19 we can work with functions $v \in \mathcal{V}_0$, see (2.27), which are smooth with singular Hessian, since they are dense in the restriction to X_v. By [FJM06, Proposition 9] there exists a displacement $u: \omega \to \mathbb{R}^2$ in $W^{2,2}(\omega; \mathbb{R}^2)$ such that

$$\nabla_s u + \frac{1}{2} \nabla v \otimes \nabla v = 0. \tag{†}$$

Fix $\delta > 0$ and, using Corollary 2.23, choose smooth functions $a \in C^\infty(\overline{\omega})$, $g \in C^\infty(\overline{\omega}; \mathbb{R}^2)$ such that

$$\| \nabla_s g + a \nabla^2 v - A_{\min} \|_{0,2}^2 < \delta,$$

where $A_{\min} \in L^\infty(\omega; \mathbb{R}_{\text{sym}}^{2\times2})$ is defined as

$$A_{\min} := \underset{A \in \mathbb{R}_{\text{sym}}^{2\times2}}{\arg\min} \int_{-1/2}^{1/2} Q_2(t, A - t \nabla^2 v + \check{B}(t)) \, \mathrm{d}t.$$

Define now the recovery sequence $(u_\theta, v_\theta)_{\theta > 0}$ with

$$u_\theta := u + \frac{1}{\sqrt{\theta}}(a \nabla v + g), \quad v_\theta := v - \frac{1}{\sqrt{\theta}} a.$$

Clearly $v_\theta = v - \theta^{-1/2} a \to v$ as $\theta \to \infty$ in $W^{2,2}(\omega)$. Furthermore

$$\sqrt{\theta} \nabla_s u_\theta = \sqrt{\theta} \nabla_s u + \nabla_s g + (\nabla a \otimes \nabla v)_s + a \nabla^2 v,$$

$$\frac{\sqrt{\theta}}{2} \nabla v_\theta \otimes \nabla v_\theta = \frac{\sqrt{\theta}}{2} \nabla v \otimes \nabla v + \frac{1}{2\sqrt{\theta}} \nabla a \otimes \nabla a - (\nabla a \otimes \nabla v)_s,$$

and[2.19]

$$-t \nabla^2 v_\theta = -t \nabla^2 v + \frac{t}{\sqrt{\theta}} \nabla^2 a,$$

2.19. Observe that this line implies that v appears with the correct scaling (i.e. none) in v_θ.

so that, using (†) and the fact that the product $\|\nabla a \otimes \nabla a\|_{0,2} = \|\nabla a\|_{0,4}^2$ is bounded we have

$$
\begin{aligned}
\mathcal{J}_{vK}^\theta(u_\theta, v_\theta) &= \frac{1}{2} \int_\omega \int_{-\frac{1}{2}}^{\frac{1}{2}} Q_2\big(t, \theta^{\frac{1}{2}} A_\theta - t\nabla^2 v_\theta + \check{B}(t)\big) \, dt \, dx' \\
&= \frac{1}{2} \int_\omega \int_{-\frac{1}{2}}^{\frac{1}{2}} Q_2(t, \nabla_s g + (a-t)\nabla^2 v + \check{B}(t)) \, dt \, dx' + \mathcal{O}\big(\theta^{-\frac{1}{2}}\big).
\end{aligned}
$$

Now subtract and add $A_{\min}$ inside Q_2 and use Cauchy's inequality (Lemma A.16) with $\varepsilon = \sqrt{\delta}$:

$$
\int_{-\frac{1}{2}}^{\frac{1}{2}} Q_2(t, \nabla_s g + a\nabla^2 v - t\nabla^2 v + \check{B}) \, dt
$$

$$
\leqslant \big(1 + \sqrt{\delta}\big) \int_{-\frac{1}{2}}^{\frac{1}{2}} Q_2(t, A_{\min} - t\nabla^2 v + \check{B}) \, dt
$$

$$
+ \frac{1}{4\sqrt{\delta}} \underbrace{\int_{-\frac{1}{2}}^{\frac{1}{2}} Q_2(t, \nabla_s g + a\nabla^2 v - A_{\min}) \, dt}_{\lesssim \|\nabla_s g + a\nabla^2 v - A_{\min}\|_{0,2}^2 < \delta}
$$

$$
= \int_{-\frac{1}{2}}^{\frac{1}{2}} Q_2(t, A_{\min} - t\nabla^2 v + \check{B}) \, dt + \mathcal{O}_{\delta\downarrow 0}\big(\delta^{\frac{1}{2}}\big).
$$

We plug this in and obtain:

$$
\begin{aligned}
\mathcal{J}_{vK}^\theta(u_\theta, v_\theta) &\leqslant \frac{1}{2} \int_\omega \int_{-\frac{1}{2}}^{\frac{1}{2}} Q_2(t, A_{\min} - t\nabla^2 v + \check{B}(t)) \, dt \, dx' \\
&\quad + \mathcal{O}\big(\theta^{-\frac{1}{2}}\big) + \mathcal{O}_{\delta\downarrow 0}\big(\delta^{\frac{1}{2}}\big) \\
&\xrightarrow{\theta\uparrow\infty} \frac{1}{2} \int_\omega \int_{-\frac{1}{2}}^{\frac{1}{2}} Q_2(t, A_{\min} - t\nabla^2 v + \check{B}(t)) \, dt \, dx' + \mathcal{O}_{\delta\downarrow 0}\big(\delta^{\frac{1}{2}}\big).
\end{aligned}
$$

The proof is concluded by letting $\delta \to 0$ and passing to a diagonal sequence. $\qquad\square$

We finish the proof of Theorem 2.7 with the lower and upper bounds for the transition from $\alpha = 3$ to $\alpha > 3$. The lack of constraints in the limit functional makes the proofs straightforward.

Theorem 2.17. (Lower bound, von Kármán to linearised von Kármán)

Let $(u_\theta, v_\theta)_{\theta>0}$ be a sequence in X_w such that $(\theta^{1/2} u_\theta, v_\theta) \to (u, v)$ in X_w as $\theta \to 0$. Then

$$\liminf_{\theta \to 0} \mathcal{I}_{vK}^\theta(u_\theta, v_\theta) \geqslant \mathcal{I}_{lvK}(u, v).$$

Proof. We may assume that $\sup_{\theta>0} \mathcal{I}_{vK}^\theta(u_\theta, v_\theta) \leqslant C$. Then by Theorem 2.14 $(\nabla v_\theta)_{\theta>0}$ is bounded in $W^{1,2}$ and by the Sobolev embedding $W^{1,2} \hookrightarrow L^4$ we have as before $\|\nabla v_\theta \otimes \nabla v_\theta\|_{0,2} = \|\nabla v_\theta\|_{0,4}^2 \leqslant C$. Consequently

$$\sqrt{\theta} A_\theta = \sqrt{\theta} \nabla_s u_\theta + \frac{\sqrt{\theta}}{2} \nabla v_\theta \otimes \nabla v_\theta \rightharpoonup \nabla_s u \quad \text{in } L^2 \text{ as } \theta \downarrow 0.$$

The functional $\mathcal{I}_{vK}^\theta$ is a Γ-limit under weak convergence, so it is w.s.l.s.c. and:

$$\liminf_{\theta \downarrow 0} \mathcal{I}_{vK}^\theta(u_\theta, v_\theta) \geqslant \frac{1}{2} \int_\omega \int_{-1/2}^{1/2} Q_2(x_3, \nabla_s u - x_3 \nabla^2 v + \check{B}(x_3)) \, dx_3 \, dx'$$
$$= \mathcal{I}_{lvK}(u, v).$$

$\square$

Theorem 2.18. (Upper bound, von Kármán to linearised von Kármán)

Let $(u, v) \in X_w$. There exists a sequence $(u_\theta, v_\theta)_{\theta>0} \subset X_w$ converging to $(u, v) \in X_w$ such that $\mathcal{I}_{vK}^\theta(u_\theta, v_\theta) \to \mathcal{I}_{lvK}(u, v)$ as $\theta \to 0$.

Proof. Define

$$u_\theta := \theta^{-1/2} u \quad \text{and} \quad v_\theta := v.$$

Clearly $(\theta^{1/2} u_\theta, v_\theta) \equiv (u, v)$ and using again $W^{1,2} \hookrightarrow L^4$ we have:

$$\sqrt{\theta} A_\theta = \nabla_s u + \frac{1}{2} \theta^{1/2} \nabla v \otimes \nabla v \xrightarrow[\theta \downarrow 0]{} \nabla_s u \quad \text{in } L^2.$$

Consequently:

$$\mathcal{I}_{vK}^\theta(u_\theta, v_\theta) = \frac{1}{2} \int_\omega \int_{-1/2}^{1/2} Q_2\left(x_3, \sqrt{\theta} A_\theta - x_3 \nabla^2 v_\theta + \check{B}(x_3)\right) dx_3 \, dx'$$
$$\xrightarrow[\theta \downarrow 0]{} \frac{1}{2} \int_\omega \int_{-1/2}^{1/2} Q_2(x_3, \nabla_s u - x_3 \nabla^2 v + \check{B}(x_3)) \, dx_3 \, dx'$$
$$= \mathcal{I}_{lvK}(u, v),$$

as stated.

$\square$

2.5 Approximation and representation theorems

A key ingredient in the proofs of the upper bounds is the density of certain smooth functions in the space where the energy is minimised. In particular, for the case $\alpha \in (2,3)$ we obtain a result proving that $W^{2,2}$ maps with singular Hessian can be approximated by smooth functions with the same property. In order to apply the results of [Sch07b] we may restrict ourselves to isometries which partition ω into finitely many so-called *bodies* and *arms* (cf. Definition 3.6). There it is shown that the set

$$\mathscr{A}_0 := \{y \in C^\infty(\overline{\omega}; \mathbb{R}^3) : y \text{ is an isometry finitely partitioning } \omega\},$$

is dense in the $W^{2,2}$-isometries. Here we show that, additionally,

$$\mathscr{V}_0 := \{v \in C^\infty(\overline{\omega}) : \exists \eta > 0 \text{ s.t. } \eta v = y_3 \text{ for some } y \in \mathscr{A}_0\} \tag{2.27}$$

is $W^{2,2}$-dense in $W_{sh}^{2,2}$.

Theorem 2.19. *Let* $\omega \subset \mathbb{R}^2$ *be a bounded, convex, Lipschitz domain. Then the set* $\mathscr{V}_0$ *is* $W^{2,2}$-*dense in* $W_{sh}^{2,2}(\omega)$. *In particular* $\det \nabla^2 v = 0$ *for all* $v \in \mathscr{V}_0$.

Proof. [2.20]

Step 1: *Approximation.*

Let $v \in W_{sh}^{2,2}(\omega)$ and $\varepsilon > 0$. By [FJM06, Theorem 10], we can find some $\tilde{v} \in W_{sh}^{2,2}(\omega) \cap W^{1,\infty}(\omega)$ s.t. $\|v - \tilde{v}\|_{2,2} < \varepsilon/2$ and, for $\eta = \eta(\varepsilon) > 0$ sufficiently small, $\|\nabla \eta \tilde{v}\|_\infty < 1/2$. One can now apply [FJM06, Theorem 7] to construct an isometry $\tilde{y} \in W^{2,2}(\omega; \mathbb{R}^3)$ whose out-of-plane component $\tilde{y}_3 = \eta \tilde{v}$. By [Sch07b, Proposition 2.3][2.21] we find a smooth $y \in \mathscr{A}_0$ such that $\|y - \tilde{y}\|_{2,2} < \varepsilon \eta/2$ and in particular $\|y_3 - \tilde{y}_3\|_{2,2} < \varepsilon \eta/2$. Setting $\psi := y_3/\eta \in \mathscr{V}_0$ we conclude

$$\|v - \psi\|_{2,2} \leqslant \|v - \tilde{v}\|_{2,2} + \|\tilde{v} - \psi\|_{2,2} < \varepsilon.$$

2.20. We are grateful to Prof. Hornung for the help provided with this proof.

2.21. It is in order to apply this theorem that we require the additional assumption (wrt. Assumption 2.1) of convexity of the domain.

Step 2: *Inclusion.*

Let $v \in \mathcal{V}_0$ with $\eta \, v = y_3$, $\eta > 0$ for some smooth isometry $y \in \mathscr{A}_0$. Recall that the second fundamental form $\mathrm{II}_{(y)}$ of any smooth isometric immersion y is singular and the identity $\nabla^2 y_j = -\mathrm{II}_{(y)} n_j$ holds for all $j \in \{1, 2, 3\}$, where $n = y_{,1} \wedge y_{,2}$.[2.22] Therefore $\det(\eta \, \nabla^2 v) = \det(-\mathrm{II}_{(y)} n_3) = 0$ and the proof is complete. $\square$

Remark 2.20. In [FJM06, Theorem 7], maps $y' \in W^{1,2}(\omega; \mathbb{R}^2)$ with $\det \nabla y' > 0$ such that $\nabla^{\mathsf{T}} y' \nabla y' + \nabla v \otimes \nabla v = I$ are built from maps $v \in W_{sh}^{2,2}(\omega)$ under the condition that $\|\nabla v\|_\infty < 1$. This additional requirement for the existence of the isometries $y = (y', v)$ is necessary and is the reason for the factor η in the definition of $\mathcal{V}_0$ (since we want to approximate any $v \in W_{sh}^{2,2}$ with arbitrarily large gradient). Indeed if we require $\det \nabla y' > 0$ then

$$0 < \det^2 \nabla y' = \det(\nabla^{\mathsf{T}} y' \nabla y') = \det(I - \nabla v \otimes \nabla v) = 1 - |\nabla v|^2,$$

and the requirement follows.[2.23]

We note that the following similar statement can be proved using the same approximation arguments and [Hor11, Theorem 1] (with the bonus of in addition holding for more general domains).

Theorem 2.21. *Let $\omega \subset \mathbb{R}^2$ be a bounded, simply connected, Lipshitz domain whose boundary contains a set $\Sigma = \overline{\Sigma} \subset \partial\omega$ with $\mathscr{H}^1(\Sigma) = 0$ such that on its complement $\partial\omega \backslash \Sigma$ the outer unit normal to ω exists and is continuous. Then the set $W_{sh}^{2,2}(\omega) \cap C^\infty(\overline{\omega})$ is $W^{2,2}$-dense in $W_{sh}^{2,2}(\omega)$.*

2.22. See [MP05, Proposition 3] for a proof for $W^{2,2}$ isometries on Lipschitz domains.

2.23. One way to see this intuitively is that it should not be possible to construct isometries from vertical displacement fields v with large gradients. A complementary view to have is that any linear map with matrix $e \otimes e$ is a projection onto the vector e, and has eigenvectors e and $e^\perp$, with respective eigenvalues $|e|^2$ and 0, as can be readily checked. This means that $|e|$ must be <1 or $I - e \otimes e$ will have one non-positive eigenvalue, hence non-positive determinant.

Once one can work with smooth functions, the essential tool for the construction of the recovery sequences for $\alpha \in (2,3)$ is the following representation theorem for maps with singular Hessian and its corollary, both following [Sch07b] very closely. A crucial component in the proof of the result in that paper is the ability to use approximations partitioning the domain in regions over which they are affine. This is in close connection to the *rigidity property* for $W^{2,2}$-isometries proved in [Pak04, Theorem II]: every point of their domain lies either on an open set or on a segment connecting the boundaries where the map is affine.

Theorem 2.22. *Let $v \in \mathcal{V}_0$ and $A \in C^\infty(\bar{\omega}; \mathbb{R}^{2 \times 2}_{\text{sym}})$ such that $A \equiv 0$ in a neighbourhood of $\{\nabla^2 v = 0\}$. There exist maps $a, g_1, g_2 \in C^\infty(\bar{\omega})$ such that $a = g_i = 0$ on $\{\nabla^2 v = 0\}$ and*

$$A = \nabla_s g + a \nabla^2 v.$$

Proof. Let $\eta > 0$, $y \in \mathcal{A}_0$ s.t. $\eta v = y_3$. Using that $\nabla^2 y_3 = -\text{II}_{(y)} n_3$ holds by virtue of y being an isometry, with $n = y_{,1} \wedge y_{,2}$ being the unit normal vector, we have that $A \equiv 0$ in a neighbourhood of $\{\text{II}_{(y)} = 0\} \cup \{n_3 = 0\}$, and

$$\{\nabla^2 v = 0\} = \{\text{II}_{(y)} = 0\} \cup \{n_3 = 0\}.$$

We can apply [Sch07b, Lemma 3.3][2.24] to y in order to obtain functions $\tilde{a}, g_1, g_2 \in C^\infty(\bar{\omega})$ s.t. $\tilde{a}, g_1, g_2 = 0$ on $\{\text{II}_{(y)} = 0\}$ and $A = \nabla_s g + \tilde{a} \text{II}_{(y)}$.

By examining the proof of this Lemma one can see that $\tilde{a}, g \equiv 0$ in a neighbourhood of $\{n_3 = 0\}$: since over bodies one has $\tilde{a}, g_1, g_2 = 0$ by construction, we need only consider arms, i.e. domains covered by a leading curve (see Definition 3.6). On these sets, if n_3 vanishes at a point then it vanishes at a whole line perpendicular to the leading curve, because the latter is orthogonal to the level sets of the gradient. Now, because $A = 0$ in a neighbourhood of this line, when solving the equations in the proof of the Lemma which determine g then $\tilde{a}$, one obtains $u_{2,s} = 0$ and $u_{2,t} = 0$, and with the boundary conditions $u_2 = 0$ then $u_1 = 0$ is a solution to the remaining equation. Hence $g = 0$ and $\tilde{a} = 0$ on these lines. Since the functions so obtained are C^∞, we can define $a := -\tilde{a} \eta / n_3$ if $n_3 \neq 0$ and $a = 0$ otherwise, and this is a smooth function such that

$$A = \nabla_s g + a \nabla^2 v. \qquad \square$$

2.24. Namely: If $y \in \mathcal{A}_0$ and $A \in C^\infty(\bar{\omega}; \mathbb{R}^{2 \times 2}_{\text{sym}})$ vanishes over a neighbourhood of $N = \{\text{II}_{(y)} = 0\}$, then there exist $\tilde{a}, g_1, g_2 \in C^\infty(\omega)$ vanishing on N such that $A = \nabla_s g + \tilde{a} \text{II}_{(y)}$.

Corollary 2.23. ([Sch07b]) *Let $v \in \mathcal{V}_0$ and define for every $x' \in \omega$*

$$A_{\min}(x') = \operatorname*{argmin}_{A \in \mathbb{R}^{2 \times 2}_{\text{sym}}} \int_{-1/2}^{1/2} Q_2(t, A - t \nabla^2 v(x') + \check{B}_s) \, dt.$$

Then $A_{\min} \in L^2(\omega; \mathbb{R}^{2 \times 2}_{\text{sym}})$ and there exist sequences of functions $a_k \in C_0^\infty(\{\nabla^2 v \neq 0\}), g_k \in C_0^\infty(\{\nabla^2 v \neq 0\}; \mathbb{R}^2)$ such that

$$\|\nabla_s g_k + a_k \nabla^2 v - A_{\min}\|_{L^2(\omega; \mathbb{R}^{2 \times 2})} \longrightarrow 0 \text{ as } k \to \infty.$$

Proof. Let $k \in \mathbb{N}$ be arbitrary. First, on the set $\{\nabla^2 v = 0\}$ we trivially have $A_{\min} \equiv A_0$ a constant matrix. Now let $A_k \in C_0^\infty(\{\nabla^2 v \neq 0\}; \mathbb{R}^{2 \times 2})$ such that

$$\|A_k - (A_{\min} - A_0)\|_{L^2(\omega; \mathbb{R}^{2 \times 2})} < \frac{1}{k}$$

and use Theorem 2.22 to pick smooth $a_k, \tilde{g}_k$ with support on $\{\nabla^2 v \neq 0\}$ and

$$A_k = \nabla_s \tilde{g}_k + a_k \nabla^2 v.$$

Set $g_k(x') = \tilde{g}_k(x') + A_0 x'$. Then:

$$\|\nabla_s g_k + a_k \nabla^2 v - A_{\min}\|_{L^2} = \|\nabla_s \tilde{g}_k + a_k \nabla^2 v - (A_{\min} - A_0)\|_{L^2} < \frac{1}{k}. \qquad \square$$

3

Properties and characterisation of minimisers

In this chapter we pursue two goals: First we develop a characterisation of minimisers for the lower range $\alpha \in (2, 3)$ of scalings in Theorem 3.1 and for the upper range $\alpha > 3$ in Theorem 3.2. The results indicate that the characteristic shapes are roughly (infinitesimal) cylinders and paraboloids respectively. Together with the interpolating property of the regime $\alpha = 3$ from Section 2.4 this leads us to conjecture that minimisers in this regime will experience a stark qualitative change for some critical (order of) magnitude $\theta_c > 0$ of the parameter. Second, we take some first steps investigating in this direction with proofs of existence, uniqueness and global minimisation for small values of θ in Section 3.2.

3.1 Optimal configurations in the linearised regimes

This first section contains the proofs of the following two theorems.

Theorem 3.1. *Up to the addition of an affine transformation, the minimisers of* $\mathcal{J}_{lKi}$ *are of the form*

$$v(x') = \frac{1}{2} x'^{\top} F x,$$

where

$$F \in \mathcal{N} := \operatorname{argmin} \{ Q_2^*(F - F_0) : F \in \mathbb{R}_{\text{sym}}^{2 \times 2}, \det F = 0 \},$$

and Q_2^*, F_0 *are given in Proposition 3.3.*

Theorem 3.2. *The minimisers of $\mathscr{J}_{lvK}$ are of the form*

$$u(x') = E_0 x' \quad \text{and} \quad v(x') = \frac{1}{2} x'^\top F_0 x',$$

where $E_0, F_0 \in \mathbb{R}^{2\times 2}_{\text{sym}}$ are constants depending on Q_2 and B which are explicitly computed in Proposition 3.8. u is unique up to an infinitesimal rigid motion and v up to the addition of an affine transformation.

We start with $\mathscr{J}_{lKi}$ and, mimicking [Sch07b, Proposition 3.5], we seek to compute a representation of the form $\overline{Q}_2^\star$ which dispenses with the minimum:

Proposition 3.3. *Let $\overline{Q}_2^\star$ be given as in (2.5):*

$$\overline{Q}_2^\star(F) := \min_{E \in \mathbb{R}^{2\times 2}} \overline{Q}_2(E, F) = \min_{E \in \mathbb{R}^{2\times 2}_{\text{sym}}} \int_{-1/2}^{1/2} Q_2(t, E + tF + \check{B}(t)) \, dt.$$

There exist a quadratic form Q_2^* strictly convex over $\mathbb{R}^{2\times 2}_{\text{sym}}$ and constants $F_0 \in \mathbb{R}^{2\times 2}, a_0 \in \mathbb{R}$, all explicitly computable in terms of the coefficients of Q_2 and B, such that:

$$\overline{Q}_2^\star(F) = Q_2^*(F - F_0) + a_0, \text{ for all } F \in \mathbb{R}^{2\times 2}_{\text{sym}}.$$

An immediate consequence, which also follows from a direct computation, is that in the homogeneous case (cf. Section 2.2.1), the linearised Kirchhoff energy reduces to a "pure bending" energy term:

Corollary 3.4. *If $Q_2(t, A) = Q_2(A)$ and $B(t) = tI$ then one has*

$$\overline{Q}_2^\star(F) = \frac{1}{12} Q_2(F + I).$$

Therefore:

$$\mathscr{J}_{lKi}(v) = \begin{cases} \frac{1}{24} \int_\omega Q_2(\nabla^2 v - I) & \text{if } v \in W^{2,2}_{sh}, \\ +\infty & \text{otherwise.} \end{cases}$$

For the proof of Proposition 3.3 we require the following:

Lemma 3.5. *Let* $t \mapsto M(t) \in \mathbb{R}^{n \times n}_{\text{sym}}$ *for* $t \in \left(-\frac{1}{2}, \frac{1}{2}\right)$ *be a measurable map such that* $M(t)$ *is positive definite for each t and define its moments*

$$M_0 := \int_{-\frac{1}{2}}^{\frac{1}{2}} M(t)\, dt, \quad M_1 := \int_{-\frac{1}{2}}^{\frac{1}{2}} t M(t)\, dt, \quad M_2 := \int_{-\frac{1}{2}}^{\frac{1}{2}} t^2 M(t)\, dt.$$

Then M_0, M_2 *and* $M^* := M_2 - M_1 M_0^{-1} M_1$ *are positive definite.*

Proof. Define first $\delta(t) := \inf_x \frac{x^\top M(t) x}{|x|^2}$, a measurable function with $\delta(t) > 0$ for every $t \in \left(-\frac{1}{2}, \frac{1}{2}\right)$. Then

$$x^\top M_0 x = x^\top \int_{-\frac{1}{2}}^{\frac{1}{2}} M(t)\, dt\, x \geqslant \int_{-\frac{1}{2}}^{\frac{1}{2}} \delta(t)\, dt\, |x|^2 = \delta_0 |x|^2,$$

with $\delta_0 > 0$. Analogously:

$$x^\top \int_{-\frac{1}{2}}^{\frac{1}{2}} t^2 M(t)\, dt\, x \geqslant \int_{-\frac{1}{2}}^{\frac{1}{2}} t^2 \delta(t)\, dt\, |x|^2 \geqslant \delta_1 |x|^2,$$

for some $\delta_1 > 0$, and this concludes the proof for M_0 and M_2. Now fix some $\Lambda \in \mathbb{R}^{2 \times 2}$ and consider the quantity

$$
\begin{aligned}
0 &\leqslant \int_{-\frac{1}{2}}^{\frac{1}{2}} \left(t M^{\frac{1}{2}}(t) - M^{\frac{1}{2}}(t) \Lambda\right)^\top \left(t M^{\frac{1}{2}}(t) - M^{\frac{1}{2}}(t) \Lambda\right) dt \\
&= \int_{-\frac{1}{2}}^{\frac{1}{2}} (tI - \Lambda) M(t) (tI - \Lambda)\, dt \\
&= M_2 - \Lambda^\top M_1 - M_1 \Lambda + \Lambda^\top M_0 \Lambda.
\end{aligned}
$$

Choosing $\Lambda = M_0^{-1} M_1$ we obtain

$$0 \leqslant M_2 - M_1 M_0^{-1} M_1,$$

and we claim that the right hand side cannot have 0 as eigenvalue and therefore the inequality is strict and the proof is concluded. If it had, say with eigenvector x_0, then

$$
\begin{aligned}
0 &= x_0^\top (M_2 - M_1 M_0^{-1} M_1) x_0 \\
&= \int_{-\frac{1}{2}}^{\frac{1}{2}} x_0^\top \left(t M^{\frac{1}{2}}(t) - M^{\frac{1}{2}}(t)\Lambda\right)^\top \left(t M^{\frac{1}{2}}(t) - M^{\frac{1}{2}}(t)\Lambda\right) x_0 \, dt \\
&= \int_{-\frac{1}{2}}^{\frac{1}{2}} \left|\left(t M^{\frac{1}{2}}(t) - M^{\frac{1}{2}}(t)\Lambda\right) x_0\right|^2 dt.
\end{aligned}
$$

So the integrand would be a.e. 0. Factoring $M^{\frac{1}{2}} > 0$ out this would mean that $(t I - \Lambda) x_0 = 0$ for a.e. t, a contradiction since Λ has at most two eigenvalues. $\qquad\square$

Proof of Proposition 3.3. Recall again definition (2.5):

$$
\overline{Q}_2^\star(F) := \min_{E \in \mathbb{R}^{2\times 2}} \int_{-\frac{1}{2}}^{\frac{1}{2}} Q_2(t, E + tF + \check{B}(t)) \, dt, \quad \text{for } F \in \mathbb{R}^{2\times 2}.
$$

Because, by Lemma A.13, $Q_2(t, F) = Q_2(t, F_s)$ we may restrict our attention to $F \in \mathbb{R}^{2\times 2}_{\mathrm{sym}}$. From now on, we identify matrices $E = (E_{ij})_{i,j=1}^2 \in \mathbb{R}^{2\times 2}_{\mathrm{sym}}$ with vectors in $\mathbb{R}^3$ via

$$
E \mapsto e := (E_{11}, E_{22}, E_{12}), \tag{3.1}
$$

and analogously $F \mapsto f$, $\check{B} \mapsto b$, $A \mapsto a$. Then, for each $t \in \left(-\frac{1}{2}, \frac{1}{2}\right)$ there exists some symmetric, positive definite matrix $M(t)$ such that for all $A \in \mathbb{R}^{2\times 2}_{\mathrm{sym}}$:

$$
Q_2(t, A) = a^\top M(t) a.
$$

Define the moments of M as in Lemma 3.5, as well as the vectors

$$
b_1 := \int_{-\frac{1}{2}}^{\frac{1}{2}} M(t) b(t) \, dt, \quad b_2 := \int_{-\frac{1}{2}}^{\frac{1}{2}} t M(t) b(t) \, dt
$$

and the scalar

$$
\beta_0 := \int_{-\frac{1}{2}}^{\frac{1}{2}} b(t)^\top M(t) b(t) \, dt.
$$

By Lemma 3.5, M_0 and M_2 are positive definite. Now, because E, F are independent of t:

$$
\begin{aligned}
\int_{-\frac{1}{2}}^{\frac{1}{2}} Q_2(t, E + tF + \check{B}(t))\, dt &= \int_{-\frac{1}{2}}^{\frac{1}{2}} \{ Q_2(t, E) + t^2\, Q_2(t, F) + Q_2(t, \check{B}(t)) \\
&\qquad + 2t\, Q_2[t, E, F] + 2\, Q_2[t, E, \check{B}(t)] \\
&\qquad + 2t\, Q_2[t, F, \check{B}(t)] \}\, dt \\
&= e^\top M_0\, e + f^\top M_2\, f + \beta_0 \\
&\qquad + 2\, e^\top M_1\, f + 2\, e \cdot b_1 + 2\, f \cdot b_2 \\
&=: q_2(e, f, b).
\end{aligned}
$$

We can then write

$$
\overline{Q}_2^{\star}(F) = \min_{e \in \mathbb{R}^3} q_2(e, f, b).
$$

Setting the derivative $\partial_e\, q_2(e, f, b) = 2\, M_0\, e + 2\, M_1\, f + 2\, b_1$ equal to zero we obtain the minimiser $e_m = -M_0^{-1}(M_1 f + b_1)$ and the expression

$$
\overline{Q}_2^{\star}(F) = e_m^\top M_0\, e_m + f^\top M_2\, f + 2\, e_m^\top M_1\, f + 2\, e_m \cdot b_1 + 2\, f \cdot b_2 + \beta_0.
$$

Substituting the value of e_m into this expression we can complete the square: for some vector $f_0 \in \mathbb{R}^3$ we have

$$
\begin{aligned}
\overline{Q}_2^{\star}(F) &= f^\top \underbrace{(M_2 - M_1 M_0^{-1} M_1)}_{=:M^*} f - 2\, b_1^\top M_0^{-1} M_1\, f + 2\, f \cdot b_2 \\
&\quad - b_1^\top M_0^{-1} b_1 + \beta_0 \\
&= (f - f_0)^\top M^* (f - f_0) - f_0^\top M^* f_0 + 2\, f^\top M^* f_0 \\
&\quad - 2\, b_1^\top M_0^{-1} M_1\, f + 2\, f \cdot b_2 - b_1^\top M_0^{-1} b_1 + \beta_0.
\end{aligned}
$$

The matrix M^* is positive definite by Lemma 3.5, hence invertible so we can choose

$$
f_0 := (M^*)^{-1}(M_1 M_0^{-1} b_1 - b_2),
$$

so that $2\, f^\top M^* f_0 - 2\, b_1^\top M_0^{-1} M_1\, f + 2\, f \cdot b_2 = 0$. Then, grouping all constants together into

$$
a_0 := -f_0^\top M^* f_0 - b_1^\top M_0^{-1} b_1 + \beta_0
$$

we obtain:

$$
\overline{Q}_2^{\star}(F) = (f - f_0)^\top M^* (f - f_0) + a_0. \tag{3.2}
$$

The result now follows if we let Q_2^* stand for the quadratic form with associated matrix M^*:

$$Q_2^*(F) := f^\top M^* f \text{ for all } F \in \mathbb{R}_{\text{sym}}^{2 \times 2}. \qquad \square$$

Proof of Corollary 3.4. With the notation of the proof of Proposition 3.3, we have $M_0 = M$, $M_1 = M \int t = 0$ and $M_2 = M \int t^2 = \frac{1}{12} M$, so $M^* = \frac{1}{12} M$. Furthermore, letting $i = (1\ 1\ 0)^\top$, we obtain $b_1 = 0$, $b_2 = \frac{1}{12} M i$ and $\beta_0 = \frac{1}{12} i^\top M i$, so $f_0 = -12 M^{-1} b_2 = -i$ and $a_0 = -f_0^\top M^* f_0 + \beta_0 = 0$. Substituting into (3.2), and because $F_0 = -I$, the statement follows. $\qquad \square$

To recapitulate, according to Theorem 2.6 and Proposition 3.3 the limit functional for $\alpha \in (2, 3)$ is

$$\mathcal{F}_{lKi}(v) = \begin{cases} \frac{1}{2} \int_\omega Q_2^*(\nabla^2 v(x') - F_0) \, dx' + \frac{a_0}{2} |\omega| & \text{if } v \in W_{sh}^{2,2}, \\ \infty & \text{otherwise,} \end{cases} \qquad (3.3)$$

where Q_2^* is a strictly convex quadratic form and F_0, a_0 are constant functions of the moments of Q_2 and B.

The next difficulty for the proof of the first main result, Theorem 3.1, lies in excluding the possibility of constructing a minimiser by piecing together functions whose Hessian belongs to the set $\mathcal{N}$, all with minimal energy but lacking a nice global structure. A careful analysis of the proof of [Sch07b, Lemma 3.3] shows that it is possible to obtain a local representation of the Hessian which shows that it must be constant and singular over ω so that minimisers are (up to an affine transformation) indeed cylindrical. In order to do this we require:

Definition 3.6. *Let $y \in W^{1,2}$ be an isometry. A connected maximal subdomain of ω where ∇y is constant and y is affine whose boundary contains more than two segments inside ω is called a **body**. A **leading curve** is a curve orthogonal to the preimages of ∇y on the open regions where ∇y is not constant, parametrised by arc-length. We define an **arm** to be a maximal subdomain $\omega(\gamma)$ which is **covered** (parametrised) by some leading curve γ as follows:*

$$\omega(\gamma) \subset \{\phi_\gamma(t, s) := \gamma(t) + s\, n(t) : s \in \mathbb{R}, t \in [0, l]\},$$

*where $n(t) = \gamma'(t)^{\perp}$. We also speak of a **covered domain**.*

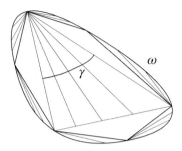

Fig. 3.1. The partition of ω into bodies and arms. ∇y is constant in the bodies (colored) and along each of the straight lines making up the arms (white).

The existence of covered domains for isometric immersions $y \in W^{1,2}$ is shown in [Pak04, Corollary 1.2].

Theorem 3.7. *Let $v \in W_{sh}^{2,2}(\omega)$ and $x_0 \in \omega$. There exists a neighbourhood U of x_0 such that, if $\nabla^2 v \neq 0$ a.e. in U, then for a suitable $\varepsilon > 0$ there exist maps $\gamma \in W^{2,2}((-\varepsilon, \varepsilon); \mathbb{R}^2)$ and $\lambda \in L^2((-\varepsilon, \varepsilon))$ such that $U \subset \{\gamma(t) + s\gamma'(t) : s \in \mathbb{R}, t \in (-\varepsilon, \varepsilon)\}$ and*

$$\nabla^2 v(\gamma(t) + s\gamma'(t)) = \frac{\lambda(t)}{1 - s\gamma''(t)} \gamma'(t) \otimes \gamma'(t) \tag{3.4}$$

if $\gamma(t) + s\gamma'(t) \in U$.

Proof. Using [FJM06, Theorem 10] take $v_k \in W^{2,2} \cap W^{1,\infty}$, $S_k \subset \omega$ such that $x_0 \in \operatorname{int} S_k$, $v_k = v$ on S_k and $\|v_k\|_{1,\infty} \leqslant C$. By scaling v_k with $\eta > 0$ we can extend ηv_k to an isometry y ([FJM06, Theorem 7]) with $\eta v_k = y_3$. Then, because y is an isometry:

$$-n_3 \operatorname{II}_{(y)} = \nabla^2 y_3 = \eta \nabla^2 v \quad \text{on } S_k.$$

Since $\nabla^2 y \neq 0$ a.e. near x_0, there is a neighbourhood U of x_0 covered by some leading curve γ, that is: $U \subset \{\gamma(t) + s\gamma'(t) : s \in \mathbb{R}, t \in (-\varepsilon, \varepsilon)\}$ and, by [Sch07b, p. 111], on U we have

$$\operatorname{II}_{(y)}(\gamma(t) + s\gamma'(t)) = \frac{\tilde{\lambda}(t)}{1 - s\gamma''(t)} \gamma'(t) \otimes \gamma'(t),$$

with $\tilde{\lambda} \in L^2$. Now, [Hor11, Proposition 1, eq. (12)] shows that $\nabla y(\gamma(t) + s\gamma'(t))$ is independent of s, hence $n_3 = (y_{,1} \wedge y_{,2})_3$ is also independent of s and we can subsume it into the function $\tilde{\lambda}$. Setting $\lambda(t) = -n_3(t)\,\tilde{\lambda}(t)/\eta$ we obtain the representation (3.4). □

Finally, we come to the proof of Theorem 3.1, which follows closely that of [Sch07b, Proposition 4.2]:

Proof of Theorem 3.1. We observe first that the set $\mathcal{N} = \mathrm{argmin}\,\{Q_2^*(F - F_0): F \in \mathbb{R}^{2\times2}_{\mathrm{sym}},\ \det F = 0\}$ is not empty because $F \mapsto Q_2^*(F - F_0)$ is non-negative and strictly convex, but it also need not consist of just one point because of the constraint. Note next that v is a minimiser of (3.3) iff $\nabla^2 v(x') \in \mathcal{N}$ for almost every $x' \in \omega$: On the one hand, every minimiser has finite energy and thus $\nabla^2 v$ must be pointwise a.e. in the set $\{F \in \mathbb{R}^{2\times2}_{\mathrm{sym}}:$ $\det F = 0\}$. On the other, any function $F: \omega \to \mathbb{R}^{2\times2}_{\mathrm{sym}}$ with $F(x') \in \mathcal{N}$ a.e. minimises the integrand in (3.3) pointwise and thus the energy.

Next we show that any two elements F, G of $\mathcal{N}$ are linearly independent. Indeed, by strict convexity we have for all $\lambda \in (0, 1)$:

$$Q_2^*(\lambda F + (1 - \lambda)\,G - F_0) < \lambda\,Q_2^*(F - F_0) + (1 - \lambda)\,Q_2^*(G - F_0).$$

Hence $\lambda F + (1 - \lambda)\,G \notin \mathcal{N}$ or else F, G would not be minimisers. Because Q_2^* attains a lower value here we must have $\det(\lambda F + (1 - \lambda)\,G) \neq 0$. But then it cannot be that $G = \rho F$ for any scalar $\rho \in \mathbb{R}$ or else it would hold that $\det(\lambda F + (1 - \lambda)\,G) = \det(\lambda F + (1 - \lambda)\,\rho F) = C \det F = 0$, a contradiction. Consequently, we have in particular $0 \notin \mathcal{N}$ unless $\mathcal{N} = \{0\}$. But in that case $\nabla^2 v \equiv 0$ and the proof would be concluded.

Let now $v \in W^{2,2}_{sh}$ be a minimiser for $\mathcal{F}_{lKi}$. Note first that ∇v cannot be constant over open sets: indeed we just saw that w.l.o.g. $0 \notin \mathcal{N}$ and consequently the possibility that $\nabla^2 v = 0$ is excluded for a minimiser. Consider then some point $x_0 \in \omega$ with a neighbourhood U where ∇v is not constant and use the representation (3.4). We have that, pointwise and over U:

$$0 \neq \nabla^2 v(t, s) = \frac{\lambda(t)}{1 - s\,\kappa(t)}\,\gamma'(t) \otimes \gamma'(t).$$

Assume now $\kappa(t_0) \neq 0$ for some t_0. By varying s we obtain distinct, linearly dependent matrices $\nabla^2 v(t_0, s_1)$ and $\nabla^2 v(t_0, s_2)$. Because $\nabla^2 v \in \mathcal{N}$ pointwise, this cannot be, so that such a t_0 does not exist: the curvature κ vanishes everywhere and consequently γ' must be constant. Analogously, λ is also constant or again we would have points at which $\nabla^2 v$ is linearly dependent. Since this holds locally around every $x' = \gamma(t) + s\,\gamma'(t)$, we deduce that $\nabla^2 v$ is constant on U and because we can cover ω in this manner, $\nabla^2 v \equiv F \in \mathcal{N}$ a.e. over ω. $\qquad\square$

Jumping now to the regime $\alpha > 3$, we proceed first to compute a simpler representation of the effective quadratic form where the misfit is absorbed into a constant.

Proposition 3.8. *Let $\alpha > 3$ and $\overline{Q}_2$ be given as in (2.4):*

$$\overline{Q}_2(E, F) := \int_{-1/2}^{1/2} Q_2(t, E + tF + \check{B}(t))\,dt.$$

There are explicitly computable constants E_0, $F_0 \in \mathbb{R}^{2\times2}_{\mathrm{sym}}$ and $c_0 \in \mathbb{R}$ depending on B and Q_2, such that

$$\overline{Q}_2(E - E_0, F - F_0) = \int_{-1/2}^{1/2} Q_2(t, E + tF) + c_0.$$

Proof. Suppose first that one can compute E_0, $F_0 \in \mathbb{R}^{2,2}_{\mathrm{sym}}$ such that the following orthogonality conditions hold

$$\begin{cases} \int_{-1/2}^{1/2} Q_2[t; \check{B}(t) - E_0 - tF_0, X]\,dt = 0, \\ \int_{-1/2}^{1/2} Q_2[t; \check{B}(t) - E_0 - tF_0, tX]\,dt = 0, \end{cases} \quad \text{for all } X \in \mathbb{R}^{2\times2}_{\mathrm{sym}}. \qquad (3.5)$$

If this is the case, then we can expand the products and after some computations arrive at the conclusion:

$$\begin{aligned} \overline{Q}_2(E - E_0, F - F_0) &= \int_{-1/2}^{1/2} Q_2(t, E - E_0 + t(F - F_0) + \check{B}(t))\,dt \\ &\overset{(3.5)}{=} \int_{-1/2}^{1/2} Q_2(t, E + tF)\,dt + c_0, \end{aligned}$$

with $c_0 = \int_{-1/2}^{1/2} Q_2(t, \check{B}(t)) - E_0 - tF_0) \, dt$.

It remains to prove the conditions (3.5). In order to do this we use the moments of M as in Lemma 3.5, as well as the vectors b_0, b_1, the scalar β and the identifications (3.1) in the proof of Proposition 3.3. With these we can write e.g. $Q_2[t; X, Y] = x^\top M(t) y$. The equations (3.5) then read

$$\begin{cases} x^\top \int_{-1/2}^{1/2} M(t)(b(t) - e_0 - tf_0) \, dt = 0, \\ x^\top \int_{-1/2}^{1/2} t M(t)(b(t) - e_0 - tf_0) \, dt = 0, \end{cases} \quad \text{for all } x \in \mathbb{R}^3,$$

which is to say that the integrals themselves vanish. Distributing the product and rearranging, this means for the first equation

$$\begin{aligned} 0 &= \int_{-1/2}^{1/2} M(t) b(t) \, dt - \int_{-1/2}^{1/2} M(t) \, dt \, e_0 - \int_{-1/2}^{1/2} t M(t) \, dt f_0 \\ &= b_1 - M_0 e_0 - M_1 f_0, \end{aligned}$$

or, with Lemma 3.5:

$$e_0 = M_0^{-1} b_1 - M_0^{-1} M_1 f_0.$$

For the second equation we have

$$0 = b_2 - M_1 e_0 - M_2 f_0,$$

and plugging in the value of e_0 and $M^* = M_2 - M_1 M_0^{-1} M_1$:

$$\begin{aligned} 0 &= b_2 - M_1 M_0^{-1} b_1 + M_1 M_0^{-1} M_1 f_0 - M_2 f_0 \\ &= b_2 - M_1 M_0^{-1} b_1 - M^* f_0. \end{aligned}$$

Again with Lemma 3.5 we can solve for f_0:

$$f_0 = (M^*)^{-1} (b_2 - M_1 M_0^{-1} b_1)$$

and consequently for e_0, concluding the proof. $\square$

Theorem 3.2 is now an inmediate consequence of the previous proposition:

Proof of Theorem 3.2. Let $u, u_0 \in W^{1,2}(\omega; \mathbb{R}^2)$ and $v, v_0 \in W^{2,2}(\omega)$ choosing u_0, v_0 such that $\nabla_s u_0 = -E_0$ and $\nabla^2 v_0 = F_0$ for the matrices E_0 and F_0 of Proposition 3.8. Then

$$
\begin{aligned}
\mathscr{J}_{lvK}(u+u_0, v+v_0) &= \frac{1}{2} \int_{\omega} \overline{Q}_2(\nabla_s u - E_0, -\nabla^2 v - F_0) \, dx' \\
&= \frac{1}{2} \int_{\omega} \int_{-1/2}^{1/2} Q_2(x_3, \nabla_s u - x_3 \nabla^2 v) \, dx_3 \, dx' + \tilde{c}_0, \\
&= \mathscr{J}_{lvK}^{B \equiv 0}(u, v) + \tilde{c}_0,
\end{aligned}
$$

where $\mathscr{J}_{lvK}^{B \equiv 0}$ denotes $\mathscr{J}_{lvK}$ with constant zero internal misfit $B \equiv 0$.

Taking infima we have

$$
\tilde{c}_0 \leqslant \inf \mathscr{J}_{lvK}(u, v) = \tilde{c}_0 + \inf \mathscr{J}_{lvK}^{B \equiv 0}(u - u_0, v - v_0),
$$

and the last summand is zero exactly when $\nabla_s u = \nabla_s u_0$ (i.e $u = u_0$ modulo infinitesimal rotations) and $\nabla^2 v = \nabla^2 v_0$ (i.e. $v = v_0$ modulo affine transformations). $\qquad \square$

3.2 The structure of minimisers for $\mathscr{J}_{vK}^{\theta}$

The second main contribution of this chapter is a first study of the properties of minimisers in the interpolating regime, "close" to the linearised von Kármán model, in the **homogeneous setting** with Q_2 independent of x_3 and linear misfit $B(t) = tI$ (see (2.13)):

$$
\mathscr{J}_{vK}^{\theta}(u, v) = \frac{\theta}{2} \int_{\omega} Q_2\left(\nabla_s u + \frac{1}{2} \nabla v \otimes \nabla v\right) dx' + \frac{1}{24} \int_{\omega} Q_2(\nabla^2 v - I) \, dx'.
$$

Natural subsequent steps along this line of work, which we do not take here, are to consider the regime $\theta \to \infty$ and to investigate the existence of the conjectured critical value θ_c, as well as to incorporate dependence on the out-of-plane component or a more general misfit.[3.1]

The first goal is to show existence and uniqueness of critical points for $\theta \ll 1$. Here the main difficulty lies in the fact that minimisers at $\theta = 0$ are not unique. Indeed,

$$
(u, v_0) \in \operatorname{argmin} \mathscr{J}_{vK}^0, \text{ for } u \text{ arbitrary and } v_0(x') = \frac{1}{2}|x'|^2 + a \cdot x' + b, \quad (3.6)
$$

[3.1]. In Chapter 4 we conduct numerical experiments supporting the conjecture that this critical value exists.

for any $a \in \mathbb{R}^2$, $b \in \mathbb{R}$, as can be readily checked by substituting. The second goal is proving that in fact these critical points are global minimisers by an application of a Taylor expansion for a carefully perturbed functional.

As before in Section 2.4, we must restrict the functions w to lie in the Banach space

$$X := X_u \times X_v,$$

with X_u, X_v as in Definition 2.12, in order to apply Korn's and Poincaré's inequalities. We give below a more natural interpretation of these requirements. The fact that this does not alter the properties of existence and uniqueness is a consequence of the arguments in Remark 2.13.

3.2.1 Existence and uniqueness for $\theta \ll 1$

Notation. *In this section, the parameter θ will be explicitly included in the arguments of the functional and differentiation is understood to be wrt. the variables $w = (u, v)$, unless otherwise stated, i.e.*

$$D\mathscr{J}_{vK}^{\theta}(u, v; \theta) = D_{u,v} \mathscr{J}_{vK}^{\theta}(u, v; \theta).$$

We are interested in the existence and uniqueness of solutions $w = (u, v)$ to the equation

$$D\mathscr{J}_{vK}^{\theta}(u, v; \theta) = 0$$

as a function of $\theta \in [0, \varepsilon)$ with $\mathscr{J}_{vK}^{\theta}$ given by (2.13). We will in fact prove the existence of a point $(u_0, v_0) \in X$ such that there exists a (locally) unique function $\phi(\theta)$, starting for $\theta = 0$ at (u_0, v_0), such that every $\phi(\theta) \in X$ is a critical point for $\mathscr{J}_{vK}^{\theta}$. However, lack of uniqueness of minimisers at $\theta = 0$ (3.6) will thwart what would be a natural application of the implicit function theorem. The problem manifests itself as a lack of injectivity of the first derivative at $(u, v) \in X$ for $\theta = 0$, which from equation (A.16) in the appendix we know to be

$$D\mathscr{J}_{vK}^{\theta}(u, v; 0)[(\varphi, \psi)] = \frac{-1}{12} \int_{\omega} Q_2[\nabla^2 v - I, \nabla^2 \psi],$$

and this vanishes at every $u \in X_u$ and the unique

$$v_0(x') = \frac{1}{2} |x'|^2 + a \cdot x' + b,$$

with $a \in \mathbb{R}^2, b \in \mathbb{R}$ such that $\int_\omega v_0 = 0$ and $\int_\omega \nabla v_0 = 0$. Because of this the equation

$$D\mathscr{I}_{vK}^\theta(u, v; \theta) = 0 \text{ in } \mathscr{L}(X, \mathbb{R})$$

cannot be uniquely solvable for $(u, v) \in X$ as a function of θ, even locally. Nevertheless, after some computations one can see that the problem in (A.16) is the presence of a leading factor θ which we can dispense with, because we may apply the implicit function theorem to the set of equivalent equations

$$\left(\frac{1}{\theta}\partial_u\right)\mathscr{I}_{vK}^\theta(u, v; \theta) = 0, \quad \partial_v \mathscr{I}_{vK}^\theta(u, v; \theta) = 0. \tag{$\diamond$}$$

These equations are equivalent to $D\mathscr{I}_{vK}^\theta(u, v; \theta) = 0$ for any $\theta > 0$ and by an application of the implicit function theorem around a specific point $(u_0, v_0; 0)$ we determine the existence of a solution function $\phi: \Theta \to U \times V$ with $[0, \varepsilon) \subset \Theta$, $\varepsilon > 0$, $U \times V \subset X$ open, $\phi(0) = (u_0, v_0)$ and $\left(\frac{1}{\theta}\partial_u, \partial_v\right)\mathscr{I}_{vK}^\theta(\phi(\theta); \theta) = 0$. Then we have $D\mathscr{I}_{vK}^\theta(\phi(\theta); \theta) = 0$ for $\theta > 0$ because of the equivalence mentioned and $D\mathscr{I}_{vK}^\theta(\phi(0); 0) = 0$ by the choice of (u_0, v_0).

We can now comment on the additional assumptions on X: without them, the derivatives in ($\diamond$) would not define an isomorphism. The fact that we need to discard constant displacements is clear, since otherwise the plate could shift away withouth incurring an energy cost. An integral condition ensures that the only allowed constant is $u \equiv 0$:

$$\int_\omega u = 0 \text{ for all } u \in X_u.$$

Given any displacement $w: \omega \to \mathbb{R}^2$ with $\nabla_s w = 0$, we can shift u by w without changing the strain: $\nabla_s u = \nabla_s(u + w)$, so one must exclude (at least) this kind of in-plane displacements from the admissible set X_u in order to even expect injectivity. Since we can always decompose the displacement gradient in *strain tensor* and *rotation tensor*, $\nabla u = \nabla_s u + \nabla_a u$, we want to exclude those u which are either constant (above) or have a vanishing strain tensor and non zero rotation tensor. Such u are called *infinitesimal rigid displacements*. They would be automatically excluded by imposing Dirichlet boundary conditions on a subset of $\partial\omega$ with positive measure, but imposing a pointwise constraint introduces serious complications for the identification of the minimisers, as we saw in the proof of Theorem 3.1. Intuitively, the reason is that it means imposing "infinitely many" constraints, thus making the function space much harder to work with. A weaker, scalar condition is sufficient:

$$\int_\omega \nabla_a u = 0 \text{ for all } u \in X_u.$$

Analogously, due to the bending energy term that will reappear in the derivative we need to exclude functions $v: \omega \to \mathbb{R}$ such that $\nabla^2 v = 0$, i.e. any affine transformation $v(x') = a\, x' + b$. The proper scalar conditions, which are those required to apply Poincaré's inequality in $W^{2,2}$ are:

$$\int_\omega v = 0 \text{ and } \int_\omega \nabla v = 0 \text{ for all } v \in X_v.$$

Finally, we will work with:

Assumptions 3.9. *The domain ω fulfils all of Assumptions 2.1 plus*

a) It is convex.
b) It is symmetric: $x' \in \omega \Leftrightarrow -x' \in \omega$.

Convexity is inessential in the current context[3.2] but it will provide the optimal Poincaré constant for the domain [PW60, eq. (1.9)] in terms of its **diameter** d:

$$\|u\|_{0,2} \leqslant \frac{d}{\pi} \|\nabla u\|_{0,2} \text{ for all } u \in W^{1,2}(\omega) \text{ such that } \int_\omega u = 0. \tag{3.7}$$

This helps in obtaining more precise estimates but it should be straightforward to dispense with it for the results of this section.

The symmetry condition provides two things that we will be using below: first, the estimate $\int_\omega |x'|^2 \leqslant d^2 |\omega|$ and second, the fact that $\int_\omega x' = 0$, hence the displacement

$$v_0(x') := \frac{1}{2}|x'|^2 - c_0,$$

with $c_0 := \frac{1}{2}(|x'|^2)_\omega$ is in X_v and it is the only minimiser of the bending energy $\frac{1}{24}\int_\omega Q_2(\nabla^2 v - I)\, \mathrm{d}x'$.

3.2. As opposed to the construction of the recovery sequence in the linearised Kirchhoff regime, Theorem 2.9.

Finally, we define the bounds

$$m := \min_{F \in \mathbb{R}^{2\times2}_{sym}, |F|=1} Q_2(F) \quad \text{and} \quad M := \max_{F \in \mathbb{R}^{2\times2}_{sym}, |F|=1} Q_2(F). \tag{3.8}$$

Theorem 3.10. *Assume the domain ω fulfills Assumptions 3.9. There exist open sets W, Θ in $X = X_u \times X_v$ and $\mathbb{R}$ respectively, with $[0, \varepsilon) \subset \Theta$, $\varepsilon > 0$, a point $u_0 \in X_u$ such that $w_0 = (u_0, v_0) \in W$ and a uniquely determined C^1 map $\phi: \Theta \to W$ such that $\phi(0) = w_0$ and*

$$D\mathcal{J}_{vK}^{\theta}(\phi(\theta); \theta) = 0 \text{ for every } \theta \in [0, \varepsilon).$$

Reciprocally, for all $w \in W$, $D\mathcal{J}_{vK}^{\theta}(w; \theta) = 0$ iff $w = \phi(\theta)$.

Proof. We first define a new set of equations to solve, then show that the second derivative of $\mathcal{J}_{vK}^{\theta}$ is one to one and then the conclusion is exactly that of the implicit function theorem, Theorem A.34. For brevity we write

$$\langle F, G \rangle := \int_{\omega} Q_2[F, G] \text{ and } \langle F \rangle := \langle F, F \rangle = \int_{\omega} Q_2(F).$$

These define a scalar product and a norm in $L^2(\omega; \mathbb{R}^{2\times2}_{sym})$ by Lemma A.16. Even though Q_2 vanishes on antisymmetric matrices, during the proof we keep track of symmetrised arguments to these functions for the sake of clarity.

Step 1: *Equivalent equations.*

From the computations leading to (A.16) we have:

$$\left(\frac{1}{\theta}\partial_u\right)\mathcal{J}_{vK}^{\theta}(u, v; \theta)[\varphi] = \langle \nabla_s u + \frac{1}{2}\nabla v \otimes \nabla v, \nabla_s \varphi \rangle,$$

and

$$\partial_v \mathcal{J}_{vK}^{\theta}(u, v; \theta)[\psi] = \theta \langle \nabla_s u + \frac{1}{2}\nabla v \otimes \nabla v, (\nabla v \otimes \nabla \psi)_{sym} \rangle$$
$$+ \frac{1}{12}\langle \nabla^2 v - I, \nabla^2 \psi \rangle.$$

We observe first that, because $\left(\frac{1}{\theta}\partial_u\right)\mathcal{J}_{vK}^{\theta}$ is independent of θ the right hand side makes sense even if $\theta = 0$. Now, on the one hand, for any fixed value of $\theta \geqslant 0$ solving the system

$$\begin{cases} \left(\frac{1}{\theta}\partial_u\right)\mathcal{J}_{vK}^{\theta}(u, v; \theta) = 0, & \text{in } \mathcal{L}(X_u, \mathbb{R}), \\ \partial_v \mathcal{J}_{vK}^{\theta}(u, v; \theta) = 0, & \text{in } \mathcal{L}(X_v, \mathbb{R}), \end{cases}$$

implies solving:

$$f(u, v; \theta)[\varphi, \psi] = 0 \text{ for every } (\varphi, \psi) \in X, \tag{3.9}$$

where $f : X \times \mathbb{R} \to \mathscr{L}(X, \mathbb{R})$ is given by

$$
\begin{aligned}
f(u, v; \theta)[\varphi, \psi] &= \langle \nabla_s u + \tfrac{1}{2} \nabla v \otimes \nabla v, \nabla_s \varphi \rangle \\
&\quad + \theta \langle \nabla_s u + \tfrac{1}{2} \nabla v \otimes \nabla v, (\nabla v \otimes \nabla \psi)_{\mathrm{sym}} \rangle \\
&\quad + \tfrac{1}{12} \langle \nabla^2 v - I, \nabla^2 \psi \rangle.
\end{aligned}
$$

On the other hand, solving $f(u, v; \theta) = 0$ for $\theta > 0$ is equivalent to solving the original problem $D\mathscr{I}_{vK}^\theta(u, v; \theta) = 0$ as we desired.

Step 2: *A zero and the derivative of f.*

Since we are interested in the behaviour around $\theta = 0$, we evaluate here and obtain

$$f(u, v; 0)[\varphi, \psi] = \langle \nabla_s u + \tfrac{1}{2} \nabla v \otimes \nabla v, \nabla_s \varphi \rangle + \tfrac{1}{12} \langle \nabla^2 v - I, \nabla^2 \psi \rangle.$$

We can compute a zero of $f(\cdot, \cdot; 0)$ by first considering the last term, which vanishes if and only if

$$v_0(x') = \tfrac{1}{2} |x'|^2 - c_0,$$

with $c_0 = \tfrac{1}{2} (|x'|^2)_\omega$. We next observe that the first term encodes the orthogonality of $\nabla_s u + \tfrac{1}{2} \nabla v_0 \otimes \nabla v_0$ to the space of symmetric gradients $\mathrm{SG}_u :=$ $\{\nabla_s \varphi : \varphi \in X_u\}$ wrt. the scalar product induced by Q_2. The $u \in X_u$ realizing this is attained by projecting onto SG_u, i.e.

$$\nabla_s u_0 = -\pi \left(\tfrac{1}{2} \nabla v_0 \otimes \nabla v_0 \right),$$

where π is the orthogonal projection defined in Lemma 3.11. We have then a point $w_0 = (u_0, v_0)$ such that

$$f(u_0, v_0; 0) = 0 \text{ in } \mathscr{L}(X, \mathbb{R}).$$

Finally, we compute $\frac{d}{d\varepsilon}\big|_{\varepsilon=0} f(u_0 + \varepsilon\,\varphi_2, v_0 + \varepsilon\,\psi_2; 0)[\varphi_1, \psi_1]$ to have the derivative of f at (u_0, v_0):

$$F_0(\varphi_2, \psi_2)[\varphi_1, \psi_1] := D_{u,v} f(u_0, v_0; 0)[(\varphi_1, \psi_1), (\varphi_2, \psi_2)]$$
$$= \langle \nabla_s \varphi_1, \nabla_s \varphi_2 + (\nabla v_0 \otimes \nabla \psi_2)_{\text{sym}} \rangle$$
$$+ \frac{1}{12}\langle \nabla^2 \psi_1, \nabla^2 \psi_2 \rangle.$$

Step 3: *The map $F_0 \colon X \to \mathcal{L}(X, \mathbb{R})$ is an isomorphism.*

Note first that the map

$$\langle (u, v), (\tilde{u}, \tilde{v}) \rangle_X := \langle \nabla_s u, \nabla_s \tilde{u} \rangle + \langle \nabla^2 v, \nabla^2 \tilde{v} \rangle$$

defines a scalar product in X, with positive-definiteness following from Korn-Poincaré's (Corollary A.22) and Poincaré's inequalities. Then we can write F_0 as

$$F_0(\varphi_2, \psi_2)[\varphi_1, \psi_1] = \langle \nabla_s \varphi_1, \nabla_s \varphi_2 + (\nabla v_0 \otimes \nabla \psi_2)_{\text{sym}} \rangle$$
$$+ \frac{1}{12}\langle \nabla^2 \psi_1, \nabla^2 \psi_2 \rangle$$
$$= \langle (\varphi_1, \psi_1), (\varphi_2 + \tilde{\pi}((\nabla v_0 \otimes \nabla \psi_2)_{\text{sym}}), \frac{1}{12}\psi_2) \rangle_X,$$

where we defined $\tilde{\pi} := \nabla_s^{-1} \circ \pi$, a continuous map from $L^2(\omega; \mathbb{R}_{\text{sym}}^{2\times2})$ to X_u as shown in the proof of Lemma 3.11, and we used that $\langle \nabla_s \varphi_1, \nabla_s \tilde{\pi}(G) \rangle = \langle \nabla_s \varphi_1, \pi(G) \rangle = \langle \nabla_s \varphi_1, \pi(G) \rangle + \langle \nabla_s \varphi_1, G - \pi(G) \rangle = \langle \nabla_s \varphi_1, G \rangle$. The Riesz representation for $F_0(\varphi_2, \psi_2)$ in $\mathcal{L}(X, \mathbb{R})$ is then $(\varphi_2 + \tilde{\pi}((\nabla v_0 \otimes \nabla \psi_2)_{\text{sym}}), \frac{1}{12}\psi_2)$ and the map

$$(\varphi_2, \psi_2) \mapsto (\varphi_2 + \tilde{\pi}((\nabla v_0 \otimes \nabla \psi_2)_{\text{sym}}), \frac{1}{12}\psi_2)$$

is clearly an isomorphism in X, with continuity for $\psi_2 \mapsto \tilde{\pi}(\nabla v_0 \otimes \nabla \psi_2)_{\text{sym}}$ following from the continuity of $\tilde{\pi}$ and the Sobolev embedding $W^{1,2} \hookrightarrow L^4$. $\qquad\square$

There were two pieces missing in the proof above: one is a Korn-Poincaré type inequality, which we leave for Appendix A.3, and the other was the definition of the orthogonal projection onto the space of symmetric gradients:

Lemma 3.11. *The map* $\pi: L^2(\omega; \mathbb{R}^{2\times2}_{\mathrm{sym}}) \to L^2(\omega; \mathbb{R}^{2\times2}_{\mathrm{sym}})$ *given by*

$$\pi(B) := \underset{A \in \mathrm{SG}_u}{\mathrm{argmin}} \int_\omega Q_2(B - A) = \underset{A \in \mathrm{SG}_u}{\mathrm{argmin}} \langle B - A \rangle_{Q_2},$$

where $\mathrm{SG}_u := \{\nabla_s u : u \in X_u\}$, *is well defined (the minimum is attained) and is a continuous projection onto* SG_u *and is orthogonal wrt. the scalar product* $\langle \cdot, \cdot \rangle_{Q_2}$. *In particular*

$$\langle B - \pi(B), \nabla_s \varphi \rangle = 0 \text{ for every } \varphi \in X_u.$$

Proof. By Lemma A.16, $L^2(\omega; \mathbb{R}^{2\times2}_{\mathrm{sym}})$ with $\langle A, B \rangle_{Q_2} := \int_\omega Q_2[A, B] \, \mathrm{d}x'$ is a Hilbert space. By the projection theorem (see e.g. [Alt12]), in order to prove the existence of a unique continuous orthogonal projection onto SG_u, it suffices to prove that it is a closed subspace of L^2, and then the projection is characterised as in the statement.

Now, the linear map $\nabla_s : X_u \to L^2(\omega; \mathbb{R}^{2\times2}_{\mathrm{sym}})$ is bounded below by Corollary A.22, hence it is injective and by the bounded inverse theorem it has a bounded linear inverse over its range, $\nabla_s^{-1} : \mathrm{SG}_u \to X_u$. Consequently the set $\mathrm{SG}_u = \nabla_s(X_u)$ is closed because X_u is. $\qquad\square$

3.2.2 Critical points are global minimisers

In addition to the previous local result, we can prove that critical points are in fact global minimisers for small non zero values of the parameter θ. We do this in two steps: close to the origin (u_0, v_0) of the branch of solutions found in the previous section, we would like to perform a Taylor expansion and use that the second differential at (u_0, v_0) is "almost" positive definite.

The key idea is to slightly modify the energy by a shift and a rescaling in order to obtain derivatives as those appearing in the equivalent equations (3.9) of Theorem 3.10, thus obtaining a positive definite second derivative. We set

$$\tilde{\mathscr{J}}^\theta_{vK}(\tilde{u}, \tilde{v}) := \mathscr{J}^\theta_{vK}\left(u_0 + \frac{\tilde{u}}{\theta}, \tilde{v}\right)$$

and then $(\tilde{u}_\theta, \tilde{v}_\theta)$ is a minimiser of $\tilde{\mathscr{J}}^\theta_{vK}$ if and only if $(u_0 + \tilde{u}_\theta/\theta, \tilde{v}_\theta)$ is a minimiser of $\mathscr{J}^\theta_{vK}$. In other words, if (u_θ, v_θ) is a minimiser of $\mathscr{J}^\theta_{vK}$, then $\tilde{u}_\theta = \theta(u_\theta - u_0)$ and $\tilde{v}_\theta = v_\theta$ minimise $\tilde{\mathscr{J}}^\theta_{vK}$.

We name $\tilde{w}_0$ the point around which we investigate the modified functional:

$$\tilde{w}_0 := (\tilde{u}_0, \tilde{v}_0) = (0, v_0). \tag{3.10}$$

Theorem 3.12. *There exists $\theta_c > 0$ such that for every $\theta \in (0, \theta_c)$, every critical point of $\tilde{\mathscr{J}}_{vK}^\theta$, and consequently of $\mathscr{J}_{vK}^\theta$, is a global minimiser.*

Proof. We proceed in three steps. First we prove that there is some $\theta_c > 0$ such that $D^2 \tilde{\mathscr{J}}_{vK}^\theta(w)$ is positive definite for all $\theta \in (0, \theta_c)$ if $\|w - \tilde{w}_0\| < \rho$ for some suitable $\rho > 0$ and $\tilde{w}_0 = (0, v_0)$ as defined in (3.10). Then we use this to determine a neighbourhood of $\tilde{w}_0$ where (local) minimisers of $\tilde{\mathscr{J}}_{vK}^\theta$ will be global by first considering points close to one such minimiser and finally those far away.

We will need the first two derivatives of $\tilde{\mathscr{J}}_{vK}^\theta$ (see Appendix A.7 for detailed, analogous computations). For the first differential we apply the chain rule to obtain $D_u \tilde{\mathscr{J}}_{vK}^\theta(u, v) = \frac{1}{\theta} D_u \mathscr{J}_{vK}^\theta\left(u_0 + \frac{u}{\theta}, v\right)$ and substitute:

$$
\begin{aligned}
D\tilde{\mathscr{J}}_{vK}^\theta(u, v)[\varphi, \psi] &= \langle \nabla_s u_0 + \tfrac{1}{\theta}\nabla_s u + \tfrac{1}{2}\nabla v \otimes \nabla v, \nabla_s \varphi \rangle \\
&\quad + \theta \langle \nabla_s u_0 + \tfrac{1}{\theta}\nabla_s u + \tfrac{1}{2}\nabla v \otimes \nabla v, (\nabla v \otimes \nabla \psi)_{\text{sym}} \rangle \\
&\quad + \tfrac{1}{12}\langle \nabla^2 v - I, \nabla^2 \psi \rangle.
\end{aligned}
$$

For the second differential we can compute another directional derivative (Lemma A.33):

$$
\begin{aligned}
\frac{d}{d\varepsilon}\Big|_{\varepsilon=0} D\tilde{\mathscr{J}}_{vK}^\theta(u + \varepsilon\,\varphi_2, v + \varepsilon\,\psi_2)[\varphi_1, \psi_1] \\
= \langle \tfrac{1}{\theta}\nabla_s \varphi_2 + (\nabla v \otimes \nabla \psi_2)_{\text{sym}}, \nabla_s \varphi_1 \rangle \\
\quad + \langle \nabla_s \varphi_2 + \theta(\nabla v \otimes \nabla \psi_2)_{\text{sym}}, (\nabla v \otimes \nabla \psi_1)_{\text{sym}} \rangle \\
\quad + \theta \langle \nabla_s u_0 + \tfrac{1}{\theta}\nabla_s u + \tfrac{1}{2}\nabla v \otimes \nabla v, (\nabla \psi_2 \otimes \nabla \psi_1)_{\text{sym}} \rangle \\
\quad + \tfrac{1}{12}\langle \nabla^2 \psi_2, \nabla^2 \psi_1 \rangle. \tag{3.11}
\end{aligned}
$$

Step 1: *There exist $\eta > 0$ and $\theta_c > 0$ s.t. $D^2 \tilde{\mathscr{J}}_{vK}^\theta(w)$ is positive definite for all $0 < \theta < \theta_c$ and all $\|w - \tilde{w}_0\|_X < \eta$.*

Observe first that it is enough to prove the statement for arguments (φ, ψ) such that $\|\varphi\|_{1,2}^2 + \|\psi\|_{2,2}^2 = 1$, i.e. we want to show that there exists some $c_0 > 0$ such that

$$D^2 \tilde{\mathscr{J}}_{vK}^\theta(w)[(\varphi, \psi), (\varphi, \psi)] \geq c_0,$$

for all θ in some interval, all w sufficiently close to $\tilde{w}_0$ and all (φ, ψ) of unit norm. Let then $\eta > 0$ be fixed and to be determined later and let $w = (u, v) \in X$ with $\|w - \tilde{w}_0\|_X < \eta$. We start by bringing terms together in (3.11):

$$
\begin{aligned}
D^2 \tilde{\mathcal{J}}_{vK}^{\theta}(w)&[(\varphi, \psi), (\varphi, \psi)] \\
&= \tfrac{1}{\theta} \langle \nabla_s \varphi + \theta (\nabla v \otimes \nabla \psi)_{\mathrm{sym}} \rangle && (a) \\
&+ \theta \langle \nabla_s u_0 + \tfrac{1}{\theta} \nabla_s u + \tfrac{1}{2} \nabla v \otimes \nabla v, (\nabla \psi \otimes \nabla \psi)_{\mathrm{sym}} \rangle && (b) \\
&+ \tfrac{1}{12} \langle \nabla^2 \psi \rangle. && (c)
\end{aligned}
$$

Given $f, g \in W^{1,2}(\omega; \mathbb{R}^2)$ we have, by the bounds (3.8) for Q_2 and Hölder's inequality (with the Sobolev embedding $W^{1,2}(\omega; \mathbb{R}^2) \hookrightarrow L^4(\omega; \mathbb{R}^2)$):

$$
\langle (f \otimes g)_{\mathrm{sym}} \rangle \lesssim \int_\omega |f \otimes g|^2 = \int_\omega |f|^2 |g|^2 \leqslant \|f\|_{0,4}^2 \|g\|_{0,4}^2 \lesssim \|f\|_{1,2}^2 \|g\|_{1,2}^2.
$$

Using this, the first and last term in $D^2 \tilde{\mathcal{J}}_{vK}^{\theta}$ can be estimated using Korn-Poincaré (Corollary A.22) and Poincaré's inequalities (3.7):

$$
\begin{aligned}
(a) &\geqslant \frac{1}{2\theta} \langle \nabla_s \varphi \rangle - \theta \langle (\nabla v \otimes \nabla \psi)_{\mathrm{sym}} \rangle \\
&\geqslant \frac{m}{2\theta} \|\nabla_s \varphi\|_{0,2}^2 - \theta M \|\nabla v \otimes \nabla \psi\|_{0,2}^2 \\
&\geqslant \frac{C_0}{\theta} \|\varphi\|_{1,2}^2 - \tilde{C}_1 \theta \|v\|_{2,2}^2 \|\psi\|_{2,2}^2 \\
&\geqslant C_0 \theta^{-1} \|\varphi\|_{1,2}^2 - C_1 \theta \|\psi\|_{2,2}^2,
\end{aligned}
$$

where in the last step we used the assumption $\|v - v_0\|_{2,2} < \eta$ to bound $\|v\|_{2,2}^2$ by some constant independent of $\eta \leqslant 1$. For the second term in $D^2 \tilde{\mathcal{J}}_{vK}^{\theta}$, use Cauchy-Schwarz (A.9) for Q_2, and the same ideas as above:

$$
\begin{aligned}
(b) &\gtrsim -\theta \left\| \nabla_s u_0 + \tfrac{1}{\theta} \nabla_s u + \tfrac{1}{2} \nabla v \otimes \nabla v \right\|_{0,2} \|(\nabla \psi \otimes \nabla \psi)_{\mathrm{sym}}\|_{0,2} \\
&\gtrsim -[\theta (\|\nabla_s u_0\|_{0,2} + \|\nabla v \otimes \nabla v\|_{0,2}) + \|\nabla_s u\|_{0,2}] \|(\nabla \psi \otimes \nabla \psi)_{\mathrm{sym}}\|_{0,2} \\
&\gtrsim -[\theta (\|u_0\|_{1,2} + \|\nabla v\|_{0,4}^2) + \|u\|_{1,2}] \|\psi\|_{0,4}^2 \\
&\gtrsim -[\theta (\|u_0\|_{1,2} + \|v\|_{2,2}^2) + \eta] \|\psi\|_{2,2}^2 \\
&\gtrsim -[\theta + \eta] \|\psi\|_{2,2}^2.
\end{aligned}
$$

Again, we used that by assumption $\|u\|_{1,2} < \eta$ and $\|v - v_0\|_{2,2} < \eta$.

Finally, we estimate the third term in $D^2 \tilde{\mathcal{J}}_{vK}^{\theta}$ with analogous arguments and obtain $(c) \geqslant C_2 \|\psi\|_{2,2}^2$. Bringing the previous computations together, we have:

$$
D^2 \tilde{\mathcal{J}}_{vK}^{\theta}(\tilde{w}) \geqslant C_0 \theta^{-1} \|\varphi\|_{1,2}^2 + (C_2 - C_1 \theta - C_3(\theta + \eta)) \|\psi\|_{2,2}^2.
$$

So there exist some θ_c and η small enough such that there exists $c_0 > 0$ with

$$D^2 \tilde{\mathcal{J}}_{vK}^{\theta}(\tilde{w})[(\varphi,\psi),(\varphi,\psi)] \geqslant c_0 \|(\varphi,\psi)\|_X,$$

for all $\tilde{w} \in X$ with $\|\tilde{w} - \tilde{w}_0\|_X < \eta$, uniformly in $\theta \in (0, \theta_c)$.

From now on, we let $\tilde{w}_\theta = (\tilde{u}_\theta, \tilde{v}_\theta)$ be a local minimiser of $\tilde{\mathcal{J}}_{vK}^{\theta}$ with

$$\|\tilde{w}_\theta - \tilde{w}_0\|_X \leqslant \eta/3 \tag{†}$$

and we prove that it is in fact a global one.

Step 2: *Estimates close to $\tilde{w}_\theta$.*

Consider first some $w \in X$ which is close to $\tilde{w}_\theta$:

$$\|w - \tilde{w}_\theta\|_X \leqslant 2\eta/3. \tag{$\star$}$$

With a Taylor expansion we see:

$$\tilde{\mathcal{J}}_{vK}^{\theta}(w) = \tilde{\mathcal{J}}_{vK}^{\theta}(\tilde{w}_\theta) + \underbrace{D\tilde{\mathcal{J}}_{vK}^{\theta}(\tilde{w}_\theta)[w - \tilde{w}_\theta]}_{=0} + \frac{1}{2}D^2\tilde{\mathcal{J}}_{vK}^{\theta}(z)[w - \tilde{w}_\theta, w - \tilde{w}_\theta],$$

where $z \in \{\alpha w + (1 - \alpha)\tilde{w}_\theta : \alpha \in [0, 1]\}$. Notice that this segment is in the η-ball around $\tilde{w}_0$ since $\|z - \tilde{w}_0\|_X \leqslant \|z - \tilde{w}_\theta\|_X + \|\tilde{w}_\theta - \tilde{w}_0\|_X \leqslant \|w - \tilde{w}_\theta\|_X + \eta/3 \leqslant \eta$, and consequently $D^2\tilde{\mathcal{J}}_{vK}^{\theta}(z) \geqslant c_0/2$.

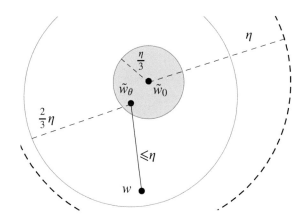

Fig. 3.2. Minimisers $\tilde{w}_\theta$ in the ▨ region are minimal at least in the ⌐ region.

Substituting in the Taylor expansion above we obtain

$$\tilde{\mathcal{J}}_{vK}^{\theta}(w) \geqslant \tilde{\mathcal{J}}_{vK}^{\theta}(\tilde{w}_{\theta}) + \frac{c_0}{4}\|w - \tilde{w}_{\theta}\|_X^2 > \tilde{\mathcal{J}}_{vK}^{\theta}(\tilde{w}_{\theta}),$$

as desired.

Step 3: *Estimates far away from $\tilde{w}_{\theta}$.*

Consider now any $w \in X$ with

$$\|w - \tilde{w}_{\theta}\|_X > 2\eta/3,$$

which in particular means that $\|w - \tilde{w}_0\|_X > \eta/3$, so we can consider two cases:

Case 1: $\|v - v_0\|_{2,2} \geqslant \eta/3$: We discard the first term in the energy, recall that $v_0(x') = |x'|^2/2 - c_0$ and use the lower bound for Q_2 in (3.8) and Poincaré's inequality (3.7):

$$\tilde{\mathcal{J}}_{vK}^{\theta}(w) \geqslant \frac{1}{24}\langle \nabla^2 v - I\rangle \geqslant \frac{m}{24}\|\nabla^2(v - v_0)\|_{0,2}^2 \geqslant C\eta^2.$$

To compare this with the energy at $\tilde{w}_0$ we add and subtract $\tilde{\mathcal{J}}_{vK}^{\theta}(\tilde{w}_0) = \frac{\theta}{2}\langle \nabla_s u_0 + \frac{1}{2}\nabla v_0 \otimes \nabla v_0\rangle$:

$$
\begin{aligned}
\tilde{\mathcal{J}}_{vK}^{\theta}(w) &\geqslant \tilde{\mathcal{J}}_{vK}^{\theta}(\tilde{w}_0) + C\frac{\eta^2}{9} - \frac{\theta}{2}\langle \nabla_s u_0 + \frac{1}{2}\nabla v_0 \otimes \nabla v_0\rangle \\
&> \tilde{\mathcal{J}}_{vK}^{\theta}(\tilde{w}_0), \qquad\qquad \text{for } \theta \text{ small enough,} \\
&\geqslant \tilde{\mathcal{J}}_{vK}^{\theta}(\tilde{w}_{\theta}),
\end{aligned}
$$

where the last line is due to the fact that $\tilde{w}_{\theta}$ minimises $\tilde{\mathcal{J}}_{vK}^{\theta}$ over the ball $B_{\frac{2}{3}\eta}(\tilde{w}_{\theta})$.

Case 2: $\|v - v_0\|_{2,2} < \eta/3$: We can estimate the energy for w as follows:

$$
\begin{aligned}
\tilde{\mathcal{J}}_{vK}^{\theta}(w) &= \frac{\theta}{2}\langle \nabla_s u_0 + \frac{1}{\theta}\nabla_s u + \frac{1}{2}\nabla v \otimes \nabla v\rangle + \frac{1}{24}\langle \nabla^2 v - I\rangle \\
&\geqslant \frac{1}{2\theta}\langle \nabla_s u\rangle + \frac{\theta}{2}\langle \nabla_s u_0 + \frac{1}{2}\nabla v \otimes \nabla v\rangle \\
&\quad + \langle \nabla_s u, \nabla_s u_0 + \frac{1}{2}\nabla v \otimes \nabla v\rangle \\
&\geqslant \left(\frac{1}{2\theta} - \varepsilon\right)\langle \nabla_s u\rangle + \left(\frac{\theta}{2} - \frac{1}{4\varepsilon}\right)\langle \nabla_s u_0 + \frac{1}{2}\nabla v \otimes \nabla v\rangle \\
&= \frac{1}{4\theta}\langle \nabla_s u\rangle - \frac{\theta}{2}\langle \nabla_s u_0 + \frac{1}{2}\nabla v \otimes \nabla v\rangle,
\end{aligned}
$$

where we chose $\varepsilon := \frac{1}{4\theta}$. Both terms may be estimated once again by a combination of the bounds (3.8) for Q_2, Sobolev's embedding $W^{1,2}(\omega) \hookrightarrow L^4(\omega)$ and Poincaré's inequality (3.7):

$$\frac{1}{4\theta} \langle \nabla_s u \rangle \gtrsim \frac{1}{\theta} \|u\|_{1,2}^2,$$

and

$$
\begin{aligned}
\frac{1}{2} \langle \nabla_s u_0 + \frac{1}{2} \nabla v \otimes \nabla v \rangle &\leqslant \langle \nabla_s u_0 \rangle + \frac{1}{2} \langle \nabla v \otimes \nabla v \rangle \\
&\lesssim \|\nabla_s u_0\|_{0,2}^2 + \|\nabla v\|_{0,4}^2 \\
&\lesssim 1 + \|v\|_{2,2}^2 \\
&\lesssim 1 + \|v - v_0\|_{2,2}^2 + \|v_0\|_{2,2}^2 \\
&\lesssim 1 + \eta^2/9.
\end{aligned}
$$

Now plug this back into the previous estimate and insert

$$\tilde{\mathscr{J}}_{vK}^{\theta}(\tilde{w}_0) = \frac{\theta}{2} \langle \nabla_s u_0 + \frac{1}{2} \nabla v_0 \otimes \nabla v_0 \rangle = \tilde{C}\theta$$

to obtain

$$
\begin{aligned}
\tilde{\mathscr{J}}_{vK}^{\theta}(w) &\gtrsim \frac{1}{\theta} \|u\|_{1,2}^2 - \theta(C + C\eta^2/9) \\
&= \frac{1}{\theta} \|u\|_{1,2}^2 - \theta(C + C\eta^2/9 + \tilde{C}) + \tilde{\mathscr{J}}_{vK}^{\theta}(\tilde{w}_0) \\
&> \tilde{\mathscr{J}}_{vK}^{\theta}(\tilde{w}_0), \qquad \text{for } \theta \text{ small enough,} \\
&\geqslant \tilde{\mathscr{J}}_{vK}^{\theta}(\tilde{w}_\theta).
\end{aligned}
$$

As above, the last line holds because $\tilde{w}_\theta$ minimises $\tilde{\mathscr{J}}_{vK}^{\theta}$ in a $\frac{2}{3}\eta$-neighbourhood of itself. $\square$

4

Discretisation of the interpolating theory

Our goal in this chapter is to study the qualitative behaviour of minimisers in the interpolating regime $\alpha = 3$ for an isotropic energy density which is homogeneous along the out-of-plane component (see Section 2.2.1). We experimentally evaluate the conjectured existence of a critical value $\theta_c > 0$ for which the symmetry of minimisers is "strongly" broken. We will not provide a full theoretical analysis, but instead adduce some empirical evidence to support the claim. To this end, we develop a simple numerical method to approximate minimisers and prove Γ-convergence to the continuous problem.

As can only be expected from a topic originating in structural mechanics, numerical methods for plate models are a vast field with a long history and as such a comprehensive review falls well beyond the scope of this thesis. However, it can be said that a significant portion of finite element approaches focus on the Euler-Lagrange equations. For von Kármán-like theories like our interpolating regime, these are transformed into an equivalent form in terms of the *Airy stress function* [HKO08, §2.6.2]. The resulting system of equations is of fourth order and can be solved with conforming C^1 elements like Argyris or specifically taylored ones. To avoid the higher number of degrees of freedom, non-conforming methods can be used instead,[4.1] but a poor choice of the discretisation can suffer from *locking*, as briefly described in Remark 4.5. Some successful classical methods employ C^0 Discrete Kirchhoff triangles (DKT), which we implement in [dB17c], but it is also possible to employ stan-

---

4.1. See [MN16b, MN16a] for particular instances of a conforming and a non-conforming method respectively, as well as reviews of recent literature.

dard Lagrange elements with penalty methods [BNRS17], as we will do. A recent line of work, upon which we heavily build in this chapter, is that of [Bar13, Bar16], where the author develops discrete gradient flows for the direct computation of (local) minimisers of non-linear Kirchhoff and von Kármán models. Γ-convergence and compactness results are also proved showing the convergence of the discrete energies to the continuous ones, as well as their respective minimisers.[4.2] Crucially, these papers use DKTs for the discretisation of the out-of-plane displacements, allowing for a representation of derivatives at nodes in the mesh which is decoupled from function values. This enables e.g. the imposition of an isometry constraint for the non-linear Kirchhoff model, but also the computation of a discrete gradient ∇_ε projecting the true gradient ∇v_ε of a discrete function v_ε into a standard piecewise P_2 space. The operator ∇_ε has good interpolation properties circumventing the lack of C^1 smoothness of DKTs which would otherwise make them unsuitable to approximate solutions in H^2.[4.3] We refer to the book [Bar15] for a systematic and mostly self-contained introduction to these methods.

4.1 Discretisation

We wish to investigate minimal energy configurations of the following functional, derived in Section 2.2.1:

$$\mathscr{I}_{vK}^\theta(u,v) = \frac{\theta}{2} \int_\omega Q_2\left(\nabla_s u + \frac{1}{2}\nabla v \otimes \nabla v\right) dx + \frac{1}{24} \int_\omega Q_2(\nabla^2 v - B) \, dx,$$

where $(u,v) \in W^{1,2}(\omega; \mathbb{R}^2) \times W^{2,2}(\omega; \mathbb{R}^2)$ and $B \in \mathbb{R}_{\text{sym}}^{2\times2}$ is constant. We assume that ω fulfils Assumptions 3.9. We implement gradient descent in a non-conforming method using C^0 linear Lagrange elements.[4.4] The

---

4.2. For a concise introduction to Γ-convergence for Galerkin discretisations and quadrature approximations of energy functionals, see [Ort04].

4.3. As an illustration of these ideas, we have implemented a pure bending model (Kirchhoff, without constraints) using both Hermite and DKT elements developed for FENiCS. See our online code repository [dB17a] for a preliminary implementation of [Bar13].

4.4. Alas, despite significant advances in the integration of DKT elements into the FENiCS framework [ABH+15] (cf. our code repositories [dB17c]), we were not able to implement in time a functioning version of the decoupled gradient flow in [Bar16] adapted to $\mathscr{I}_{vK}^\theta$.

first step is to transform the problem into one of constrained minimisation reducing the order of the elements required.

Problem 4.1. *Find minimisers of*

$$J^{\theta}(u,z) = \frac{\theta}{2} \int_{\omega} Q_2\left(\nabla_s u + \frac{1}{2} z \otimes z\right) dx + \frac{1}{24} \int_{\omega} Q_2(\nabla z - B) dx, \qquad (4.1)$$

with $B \in \mathbb{R}_{sym}^{2 \times 2}$ *a constant,* $u, z \in W^{1,2}(\omega; \mathbb{R}^2)$ *and*

$$z \in Z := \{\zeta \in W^{1,2}(\omega; \mathbb{R}^2) : \operatorname{curl} \zeta = 0\}.$$

If $z \notin Z$, *then we set* $J^{\theta}(u,z) = +\infty$.

We can now use H^1-conforming elements but, for simplicity of implementation, instead of adding the constraint into the discrete spaces to obtain a truly conforming discretisation, we add a penalty term $\mu_{\varepsilon} \|\operatorname{curl} z_{\varepsilon}\|^2$ to ensure that the solutions z_{ε} are close to gradients.

Assume from now on that ω is a polygonal domain. For fixed $\varepsilon > 0$, introduce a **quasi-uniform** triangulation $\mathscr{T}_{\varepsilon}$ of ω with triangles T of uniformly bounded diameter $c^{-1} \varepsilon \leqslant \varepsilon_T \leqslant c \varepsilon$ for some $c > 0$ and all $\varepsilon > 0$ and $T \in \mathscr{T}_{\varepsilon}$.[4.5] Such a mesh is in particular said to be, in virtue of the uniform upper bound, **shape-regular**. We denote by $\mathscr{N}_{\varepsilon}$ the set of all nodes of the triangulation. Define V_{ε} to be the standard piecewise affine, globally continuous Lagrange P_1 finite element space $\mathscr{S}^1(\mathscr{T}_{\varepsilon})$ in two dimensions:[4.6]

$$V_{\varepsilon} := \{v_{\varepsilon} \in C(\overline{\omega}; \mathbb{R}^2) : v_{\varepsilon|T} \in P_1(T)^2 \text{ for all } T \in \mathscr{T}_{\varepsilon}\}.$$

Quadrature rules will be chosen to be exact for this polynomial degree and the first integrand in the energy interpolated for this to apply by means of the **interpolated quadratic form**

$$Q_2^{\varepsilon} := \hat{I}_{\varepsilon} \circ Q_2.$$

4.5. Note that this does not allow for arbitrary local refinements or **grading** (a different scaling of simplices along different directions as $\varepsilon \to 0$), but the fact that this is not optimal is not of concern here.

4.6. Because functions in this space are Lipschitz over each compact triangle, they are globally Lipschitz and $V_{\varepsilon} \subset W^{1,2}(\omega; \mathbb{R}^2)$. However $V_{\varepsilon} \not\subset Z$ of Problem 4.1 since we do not impose any constraints, so the method will not be conforming.

This is defined (with a slight abuse of notation) component-wise using the **element-wise nodal interpolant** $\hat{I}_\varepsilon$, defined for functions $v \in L^\infty(\omega)$ such that $v_{|T} \in C(\overline{T})$ for all $T \in \mathscr{T}_\varepsilon$ as

$$\hat{I}_\varepsilon(v) := \sum_{T \in \mathscr{T}_\varepsilon} \sum_{z \in \mathscr{N}_\varepsilon \cap T} v_{|T}(z)\, \varphi_{z|T}, \tag{4.2}$$

where $\varphi_{z|T}$ is the truncation by zero outside T of the global basis function $\varphi_z \in \mathscr{S}^1$. Because this is a linear combination of truncated global basis functions, the range of $\hat{I}_\varepsilon$ is the space $\hat{\mathscr{S}}^1(\mathscr{T}_\varepsilon)$ of discontinuous, piecewise affine Lagrange elements.

In cases where the function to be interpolated is continuous, the element-wise nodal interpolant coincides with the **standard nodal interpolant** into the space $\mathscr{S}^1$ of globally continuous, piecewise affine functions, which is defined as

$$I_\varepsilon(v) := \sum_{z \in \mathscr{N}_\varepsilon} v(z)\, \varphi_z. \tag{4.3}$$

Notice that the shape functions φ_z are not truncated. In order to control the error incurred by the interpolation, we will use the following standard result:

Lemma 4.1. (Nodal interpolation estimates) *Let $v \in W^{2,p}(\overline{\omega}), 1 < p \leqslant \infty$ and let $\mathscr{T}_\varepsilon$ be a shape-regular triangulation of the polygonal domain $\overline{\omega}$. Let I_ε be the standard nodal interpolation operator onto P_1 (4.3). Then there exists some $C > 0$ independent of ε such that for every $0 \leqslant r \leqslant 2$:*

$$|I_\varepsilon(v) - v|_{r,p} \leqslant C\, \varepsilon^{2-r}\, \|D^2 v\|_{0,p},$$

where $|\cdot|$ denotes the standard seminorm.

Proof. See [GRS07, Theorem 4.28] or [BS08, (4.4.4)]. □

When working with discontinuous functions in $\hat{\mathscr{S}}^1$, we will use the following local result. This follows from Lemma 4.1 or can be shown directly, e.g. in [Bar15, Proposition 3.1].

Lemma 4.2. (Local interpolation estimate) *Let $T \in \mathscr{T}_\varepsilon$ and $v \in C^1(\overline{T})$. If $\hat{I}_\varepsilon$ is the element-wise nodal interpolant (4.2), then*

$$\|v - \hat{I}_\varepsilon(v)\|_{0,p,T} \leqslant C\, \varepsilon\, \|Dv\|_{0,p,T}.$$

The goal is to solve:

Problem 4.2. *Let* $\mu_\varepsilon > 0$. *Compute minimisers of the discrete energy*

$$J_\varepsilon^\theta(u_\varepsilon, z_\varepsilon) = \frac{\theta}{2} \int_\omega Q_2^\varepsilon \left(\nabla_s u_\varepsilon + \frac{1}{2} z_\varepsilon \otimes z_\varepsilon \right) dx + \frac{1}{24} \int_\omega Q_2(\nabla z_\varepsilon - B) \, dx$$

$$+ \mu_\varepsilon \int_\omega |\mathrm{curl}\, z_\varepsilon|^2 \, dx, \tag{4.4}$$

for $(u_\varepsilon, z_\varepsilon) \in V_\varepsilon^2$.

As usual, if $(u_\varepsilon, z_\varepsilon) \in W^{1,2}(\omega; \mathbb{R}^2)^2 \backslash V_\varepsilon^2$, we set $J_\varepsilon^\theta(u_\varepsilon, z_\varepsilon) = +\infty$.

Remark 4.3. (*Scaling of the constants*) The penalty $\mu_\varepsilon = \mu(\varepsilon)$ needs to explode as $\varepsilon \to 0$ in order for the functionals to Γ-converge (Theorem 4.8). However, large penalties negatively affect the condition number of the system, so that an adequate choice for μ_ε, dependent on the mesh size ε, is required [GRS07, p.416]. We have not explictly investigated how this requirement interacts with the Γ-convergence of the functionals, but in our proof we require only that $\mu_\varepsilon \to \infty$ not faster than ε^{-2}. In the implementation we use $\mu_\varepsilon = 1/\varepsilon^{-1}$. Analogously, large values of the Lamé constants have a similar effect and therefore hinder convergence, so one needs to scale them to the order of the problem.

Remark 4.4. (*Automatic fulfilment of the constraint*) Experiments seemed to indicate that under some circumstances, in particular not too unfavourable initialisations, one can set $\mu = 0$ and still obtain minimisers with vanishing curl. However, if this is done, when θ is increased and approaches the (conjectured) critical value, long energy plateaus are traversed after which large, markedly non physical deformations take place. Because the constraint is part of the discretisation, we did not further investigate this phenomenon.

Remark 4.5. (*Common issues with finite element methods for plates*) Discretisations for lower dimensional theories can face complications due to the infamous *locking phenomena*. In a nutshell, these mean that as the thickness of the plate tends to zero, discrete solutions "lock" to stiff states of lower, or even vanishing, bending or shearing than the analytic ones.[4.7] Another instance of unexpected behaviour is known as the

---

4.7. We refer to [BS92] for a first rigorous definition of locking, to [Pra01, Chapters 5 and 6] for detailed computations highlighting the issues with linear elements in the context of Timoshenko beams and to the thesis [Qua12] for a thorough and detailed analysis of locking in shell models.

Babuška paradox [BP90], again a failure to converge as expected, which can happen in e.g. the Kirchhoff model when both vertical and tangential displacements are fixed at the boundaries of a polygonal domain: these so-called "hard" support constraints are not enforced in the same manner as in the continuous model because of the approximated domain.

There are two potential sources of locking in our setting: the penalty term μ_ε, which is akin to the shear strain in Timoshenko beams, and θ. We have not obtained any a priori bounds on the error in this work, but a rigorous treatment of the problem would require estimates which are uniform in these parameters as the mesh diameter goes to zero. For the regimes studied and the geometries considered we have found the issue to be of moderate practical relevance, but it does manifest itself e.g. with more complicated domains or higher values of θ.

Finally, our simulations will not suffer from Babuška's paradox because we do not prescribe boundary conditions.

4.2 Γ-convergence of the discrete energies

The first step in the proof that $J_\varepsilon^\theta \overset{\Gamma}{\to} J^\theta$ is dispensing with the interpolation operators for numerical integration: due to the good properties of $\hat{I}_\varepsilon$, we can assume that we work with the true integrals $\int Q_2$ instead of $\int Q_2^\varepsilon$.

Lemma 4.6. (Numerical integration) *Let* $u_\varepsilon, z_\varepsilon \in W^{1,2}(\omega; \mathbb{R}^2)$ *be uniformly bounded in* $W^{1,2}$ *and let* $Q_2^\varepsilon = \hat{I}_\varepsilon \circ Q_2$ *as above. Let* $A_\varepsilon := \nabla_s u_\varepsilon + \frac{1}{2} z_\varepsilon \otimes z_\varepsilon$. *Then, as* $\varepsilon \to 0$:

$$\|Q_2^\varepsilon(A_\varepsilon) - Q_2(A_\varepsilon)\|_{0,1} \to 0.$$

Proof. By the local interpolation estimate Lemma 4.2:

$$\int_\omega |Q_2^\varepsilon(A_\varepsilon) - Q_2(A_\varepsilon)| \, \mathrm{d}x \lesssim \varepsilon \sum_{T \in \mathcal{T}_\varepsilon} \int_T |DQ_2(A_\varepsilon)| \, \mathrm{d}x$$

$$\lesssim \varepsilon \sum_{T \in \mathcal{T}_\varepsilon} \int_T \sum_{i=1}^2 |\partial_i Q_2(A_\varepsilon)| \, \mathrm{d}x.$$

For simplicity we work with the partial derivatives for $i \in \{1, 2\}$. Note first that, pointwise:

$$|\partial_i Q_2(A_\varepsilon)| = 2 |Q_2[A_\varepsilon, \partial_i A_\varepsilon]| = 2 q_{jklm}(A_\varepsilon)_{jk}(\partial_i A_\varepsilon)_{lm} \lesssim |A_\varepsilon||\partial_i A_\varepsilon|,$$

where $q_{jklm} \in \mathbb{R}$ are the coefficients of Q_2. Consequently, by Cauchy-Schwarz, first for integrals, then for sums:

$$\varepsilon \sum_{T \in \mathscr{T}_\varepsilon} \int_T |\partial_i Q_2(A_\varepsilon)| \, dx \lesssim \varepsilon \sum_{T \in \mathscr{T}_\varepsilon} \int_T |A_\varepsilon| |\partial_i A_\varepsilon| \, dx$$

$$\lesssim \varepsilon \sum_{T \in \mathscr{T}_\varepsilon} \left(\int_T |A_\varepsilon|^2 \, dx \right)^{1/2} \left(\int_T |\partial_i A_\varepsilon|^2 \, dx \right)^{1/2}$$

$$\lesssim \varepsilon \left(\sum_{T \in \mathscr{T}_\varepsilon} \int_T |A_\varepsilon|^2 \, dx \right)^{1/2} \left(\sum_{T \in \mathscr{T}_\varepsilon} \int_T |\partial_i A_\varepsilon|^2 \, dx \right)^{1/2}.$$

Now, the first term is simply $\|A_\varepsilon\|_{0,2,\omega}$ which is uniformly bounded since $\|z_\varepsilon \otimes z_\varepsilon\|_{0,2} = \|z_\varepsilon\|_{0,4}^2 \lesssim \|z_\varepsilon\|_{1,2}^2$, and for the second we use that $\nabla_s u_\varepsilon$ is piecewise constant so that

$$|\partial_i A_\varepsilon|^2 = |z_\varepsilon \otimes \partial_i z_\varepsilon + \partial_i z_\varepsilon \otimes z_\varepsilon|^2 \lesssim |z_\varepsilon|^2 |\partial_i z_\varepsilon|^2$$

and

$$\sum_{T \in \mathscr{T}_\varepsilon} \int_T |\partial_i A_\varepsilon|^2 \, dx \lesssim \sum_{T \in \mathscr{T}_\varepsilon} \int_T |z_\varepsilon|^2 |\partial_i z_\varepsilon|^2 \, dx \leq \sum_{T \in \mathscr{T}_\varepsilon} \|z_\varepsilon\|_{0,\infty,T}^2 \|\partial_i z_\varepsilon\|_{0,2,T}^2.$$

A standard inverse estimate (see e.g. [BS08, Theorem 4.5.11]) provides the bound

$$\max_{T \in \mathscr{T}_\varepsilon} \|z_\varepsilon\|_{0,\infty,T} \lesssim \varepsilon^{-1/2} \left(\sum_{T \in \mathscr{T}_\varepsilon} \|z_\varepsilon\|_{0,4,T}^4 \right)^{1/4}.$$

We plug this into the preceding computation to obtain

$$\sum_{T \in \mathscr{T}_\varepsilon} \int_T |\partial_i A_\varepsilon|^2 \, dx \lesssim \varepsilon^{-1} \left(\sum_{T \in \mathscr{T}_\varepsilon} \|z_\varepsilon\|_{0,4,T}^4 \right)^{1/2} \sum_{T \in \mathscr{T}_\varepsilon} \|\partial_i z_\varepsilon\|_{0,2,T}^2$$

$$= \varepsilon^{-1} \|z_\varepsilon\|_{0,4,\omega}^2 \|\partial_i z_\varepsilon\|_{0,2,\omega}^2.$$

The last two norms being uniformly bounded, we conclude:

$$\int_\omega |Q_2^\varepsilon(A_\varepsilon) - Q_2(A_\varepsilon)| \, dx \lesssim \sum_{i=1}^2 \sum_{T \in \mathscr{T}_\varepsilon} \varepsilon \int_T |\partial_i Q_2(A_\varepsilon)| \, dx \lesssim \varepsilon^{1/2} \to 0. \qquad \square$$

The second step is, as usual, to ensure that we can focus on smooth functions for simplicity in the construction of the upper bound:

Lemma 4.7. *Assume ω is convex. The set $C^\infty(\overline{\omega}) \cap Z$ is $W^{1,2}$-dense in $Z := \{z \in W^{1,2}(\omega; \mathbb{R}^2): \text{curl } z = 0\}$.*

Proof. Observe that $Z = \{\nabla v: v \in W^{2,2}(\omega)\}$. Now take any $z \in Z$ with $z = \nabla v$ and let $\varepsilon > 0$. Because $C^\infty(\overline{\omega})$ is dense in $W^{2,2}(\omega)$, we can find a smooth φ with $\|\varphi - v\|_{2,2} < \varepsilon$. But then

$$\varepsilon > \|\nabla\varphi - \nabla v\|_{1,2} = \|\nabla\varphi - z\|_{1,2}$$

and the function $\nabla\varphi \in C^\infty(\overline{\omega}) \cap Z$. □

Theorem 4.8. *Let J^θ, J^θ_ε be given by (4.1) and (4.4) respectively. Let $\mu_\varepsilon \to \infty$ but $\mu_\varepsilon = o(\varepsilon^{-2})$ as $\varepsilon \to 0$. Then $J^\theta_\varepsilon \overset{\Gamma}{\to} J^\theta$ as $\varepsilon \to 0$ wrt. weak convergence in $W^{1,2}$.*

Proof. Because of Lemma 4.6 we can substitute Q_2 for Q_2^ε in J^θ_ε. Also, by Lemmas 4.7 and A.27 it is enough to consider smooth functions for the upper bound. Set

$$A := \nabla_s u + \frac{1}{2} z \otimes z \quad \text{and} \quad A_\varepsilon := \nabla_s u_\varepsilon + \frac{1}{2} z_\varepsilon \otimes z_\varepsilon.$$

Step 1: *Upper bound.*

Let $(u, z) \in W^{1,2}(\omega; \mathbb{R}^2) \times Z$ be C^∞ up to the boundary and define $u_\varepsilon := I_\varepsilon(u)$, $z_\varepsilon := I_\varepsilon(z)$, where I_ε is the nodal interpolant of (4.3). Note that because u and z are smooth, we can apply the standard interpolation estimates of Lemma 4.1 to show strong convergence in $W^{1,2}$ of these sequences towards u and z. By (2.26), we know that $z_\varepsilon \otimes z_\varepsilon \to z \otimes z$ in L^2, so we have that $A_\varepsilon \to A$ in L^2 and by Cauchy's inequality for Q_2, Lemma A.16, for fixed $\delta > 0$ we have

$$\int_\omega |Q_2(A_\varepsilon) - Q_2(A)| = \int_\omega Q_2[A_\varepsilon + A, A_\varepsilon - A]$$

$$\leqslant \delta \int_\omega Q_2(A_\varepsilon + A) + \frac{1}{4\delta} \int_\omega Q_2(A_\varepsilon - A)$$

$$\lesssim \delta(\|A_\varepsilon\|^2_{0,2} + \|A\|^2_{0,2}) + \frac{1}{\delta}\|A_\varepsilon - A\|^2_{0,2},$$

where we also used the elementary bound for Q_2 in Proposition A.11. We can always choose $\varepsilon_0 = \varepsilon_0(\delta)$ such that $\|A_\varepsilon - A\|_{0,2} \lesssim \delta^2$ for all $\varepsilon \leqslant \varepsilon_0$ and consequently: $\int_\omega |Q_2(A_\varepsilon) - Q_2(A)| \lesssim \delta$. Since this can be done for all $\delta > 0$, we obtain

$$\lim_{\varepsilon \to 0} \int_\omega Q_2\left(\nabla_s u_\varepsilon + \tfrac{1}{2} z_\varepsilon \otimes z_\varepsilon\right) = \int_\omega Q_2\left(\nabla_s u + \tfrac{1}{2} z \otimes z\right).$$

An analogous computation shows

$$\lim_{\varepsilon \to 0} \int_\omega Q_2(\nabla z_\varepsilon - B) = \int_\omega Q_2(\nabla z - B).$$

Finally, by the same interpolation estimate above and the assumption on μ_ε we have that $\mu_\varepsilon \|\mathrm{curl}\,(\hat{I}_\varepsilon(z) - z)\|_{0,2}^2 = o(1)$ as $\varepsilon \to 0$, and consequently

$$J_\varepsilon^\theta(u_\varepsilon, z_\varepsilon) \to J^\theta(u, z).$$

Step 2: *Lower bound, finite case.*

Let $u_\varepsilon, z_\varepsilon \in V_\varepsilon \subset W^{1,2}$ with $u_\varepsilon \rightharpoonup u, z_\varepsilon \rightharpoonup z$ weakly in $W^{1,2}$ to $u \in W^{1,2}(\omega; \mathbb{R}^2)$, $z \in Z$. By (2.26), we know that $z_\varepsilon \otimes z_\varepsilon \to z \otimes z$ in L^2, so we have that $A_\varepsilon \rightharpoonup A$ in L^2. Analogously $\nabla z_\varepsilon \rightharpoonup \nabla z$ and $\mathrm{curl}\, z_\varepsilon \rightharpoonup \mathrm{curl}\, z$ also in L^2. Dropping the (non-negative) curl term in J_ε^θ and by the w.s.l.s.c. of all integrands involved (Q_2 being finite and convex [Dac07]),

$$\liminf_{\varepsilon \to 0} J_\varepsilon^\theta(u_\varepsilon, z_\varepsilon) \geqslant \frac{\theta}{2} \int_\omega Q_2\left(\nabla_s u + \tfrac{1}{2} z \otimes z\right) dx + \frac{1}{24} \int_\omega Q_2(\nabla z - B)\,dx$$
$$= J^\theta(u, z).$$

Step 3: *Lower bound, infinite case.*

Let $u_\varepsilon, z_\varepsilon \in V_\varepsilon$ with $u_\varepsilon \rightharpoonup u, z_\varepsilon \rightharpoonup z$ weakly in $W^{1,2}$ to $u \in W^{1,2}(\omega; \mathbb{R}^2)$, $z \in W^{1,2} \setminus Z$. Then, $\mathrm{curl}\, z \neq 0$ on a set of positive measure and $J^\theta(u, z) = +\infty$. We need to show that the energy $J_\varepsilon^\theta(u_\varepsilon, z_\varepsilon)$ diverges.

Suppose that there exists a subsequence, not relabeled, such that $J_\varepsilon^\theta(u_\varepsilon, z_\varepsilon) \leqslant C$. Then $\mu_\varepsilon \int_\omega |\mathrm{curl}\, z_\varepsilon|^2\, dx \leqslant C$ and $\|\mathrm{curl}\, z_\varepsilon\|_{0,2} \to 0$ as $\varepsilon \to 0$. But because $z_\varepsilon \rightharpoonup z$ in $W^{1,2}$, we have $\mathrm{curl}\, z_\varepsilon \rightharpoonup \mathrm{curl}\, z$ in L^2, a contradiction by uniqueness of the weak limits. $\qquad\square$

The final ingredient of this subsection is a proof that sequences with bounded energy are (weakly) precompact. The fundamental theorem of Γ-convergence, Lemma A.29, then shows convergence of global minimisers. In order for this to work, we need to assume conditions in the space which provide Korn and Poincaré inequalities. We can do this using functions with zero mean, zero mean of the gradient or zero mean of the antisymmetric gradient, as we do in Section 2.4 and Chapter 3, but including these conditions in the discrete spaces is not entirely trivial. Because the energies are invariant under the transformations which are factored out by taking quotient spaces as described in the sections mentioned, it is enough for our purposes to claim compactness modulo these transformations and to exclude them in the implementation via projected gradient descent.[4.8]

Theorem 4.9. (Compactness) *Let* $(u_\varepsilon, z_\varepsilon)_{\varepsilon > 0}$ *be a sequence in* $(V_\varepsilon \cap X_u)^2$ *with bounded energy, where* X_u *is defined by (2.12) in page 56. Then there exist* $u \in W^{1,2}, z \in Z$ *such that* $u_\varepsilon \rightharpoonup u$ *and* $z_\varepsilon \rightharpoonup z$ *in* $W^{1,2}$.

Proof. As above, let $A_\varepsilon := \nabla_s u_\varepsilon + \frac{1}{2} z_\varepsilon \otimes z_\varepsilon$, and set $\|F\|_{Q_2}^2 := \int Q_2(F)$. Note that we cannot use Lemma 4.6 to substitute Q_2 for Q_2^ε since we do not have uniform bounds in $W^{1,2}$ by assumption, so we work directly with J_ε^θ.

We begin by observing that by the properties of Q_2 in Lemma A.16 and all terms being non-negative, the sequence having bounded energy implies:

$$\|\nabla z_\varepsilon - B\|_{0,2}^2 \lesssim \|\nabla z_\varepsilon - B\|_{Q_2}^2 \leqslant J_\varepsilon^\theta(u_\varepsilon, z_\varepsilon) \leqslant C,$$

and consequently, by Poincaré's inequality:

$$\|z_\varepsilon\|_{1,2} \leqslant C. \tag{$\star$}$$

We have then a subsequence (not relabeled) weakly converging in $W^{1,2}$ to some $z \in W^{1,2}$. In particular $\nabla z_\varepsilon \rightharpoonup \nabla z$ and $\operatorname{curl} z_\varepsilon \rightharpoonup \operatorname{curl} z$ in L^2. But also

$$\mu_\varepsilon \|\operatorname{curl} z_\varepsilon\|_{0,2}^2 \leqslant C \Rightarrow \operatorname{curl} z_\varepsilon \to 0 \text{ in } L^2,$$

and therefore $\operatorname{curl} z = 0$, i.e. $z \in Z$.

Now, for the sequence u_ε we must work with Q_2^ε instead. First write

$$\begin{aligned}
Q_2(\nabla_s u_\varepsilon) &= Q_2\left(\nabla_s u_\varepsilon + \frac{1}{2} z_\varepsilon \otimes z_\varepsilon - \frac{1}{2} z_\varepsilon \otimes z_\varepsilon\right) \\
&\leqslant 2 Q_2(A_\varepsilon) + \frac{1}{2} Q_2(z_\varepsilon \otimes z_\varepsilon) \\
&\lesssim Q_2(A_\varepsilon) + |z_\varepsilon|^4.
\end{aligned}$$

4.8. I.e. one takes one step in the direction of the gradient, then projects the new point onto the space of admissible functions and repeats until convergence.

Since this applies pointwise, after (local) interpolation the estimate still holds:

$$Q_2^\varepsilon(\nabla_s u_\varepsilon) \lesssim Q_2^\varepsilon(A_\varepsilon) + \hat{I}_\varepsilon(|z_\varepsilon|^4).$$

Note now that because $\nabla_s u_\varepsilon$ is piecewise constant, $Q_2^\varepsilon(\nabla_s u_\varepsilon) = Q_2(\nabla_s u_\varepsilon)$, so

$$
\begin{aligned}
\int_\omega Q_2(\nabla_s u_\varepsilon) &= \int_\omega Q_2^\varepsilon(\nabla_s u_\varepsilon) \\
&\lesssim \int_\omega Q_2^\varepsilon(A_\varepsilon) + \hat{I}_\varepsilon(|z_\varepsilon|^4) \\
&\lesssim J_\varepsilon^\theta(u_\varepsilon, z_\varepsilon) + \int_\omega \hat{I}_\varepsilon(|z_\varepsilon|^4).
\end{aligned}
$$

We claim now that $\|\hat{I}_\varepsilon(|z_\varepsilon|^4) - |z_\varepsilon|^4\|_{0,1} = \mathcal{O}(\varepsilon)$. Indeed, by the local interpolation estimate (Lemma 4.2) and Hölder's inequality for integrals and for sums:

$$
\begin{aligned}
\int_\omega |\hat{I}_\varepsilon(|z_\varepsilon|^4) - |z_\varepsilon|^4| &\lesssim \varepsilon \sum_{T \in \mathscr{T}_\varepsilon} \int_T |\nabla |z_\varepsilon|^4| \\
&\lesssim \varepsilon \sum_{T \in \mathscr{T}_\varepsilon} \int_T |z_\varepsilon|^3 |\nabla z_\varepsilon| \\
&\lesssim \varepsilon \sum_{T \in \mathscr{T}_\varepsilon} \|z_\varepsilon\|_{0,6,T}^3 \|\nabla z_\varepsilon\|_{0,2,T} \\
&\lesssim \varepsilon \left(\sum_{T \in \mathscr{T}_\varepsilon} \|z_\varepsilon\|_{0,6,T}^6 \right)^{1/2} \left(\sum_{T \in \mathscr{T}_\varepsilon} \|\nabla z_\varepsilon\|_{0,2,T}^2 \right)^{1/2} \\
&\lesssim \varepsilon \|z_\varepsilon\|_{0,6,\omega}^3 \|\nabla z_\varepsilon\|_{0,2,\omega},
\end{aligned}
$$

and this goes to zero as $\varepsilon \to 0$ by ($\star$). But then $\int_\omega \hat{I}_\varepsilon(|z_\varepsilon|^4) \leqslant C$ and by Corollary A.22, the Sobolev embedding $W^{1,2} \hookrightarrow L^4$ and the previous bound, we have

$$\|u_\varepsilon\|_{1,2}^2 \lesssim \|\nabla_s u_\varepsilon\|_{0,2}^2 \lesssim \|\nabla_s u_\varepsilon\|_{Q_2}^2 \lesssim J_\varepsilon^\theta(u_\varepsilon, z_\varepsilon) + C \leqslant C.$$

The sequence $(u_\varepsilon)_{\varepsilon>0}$ is therefore also weakly precompact in $W^{1,2}(\omega; \mathbb{R}^2)$ and the proof is complete. $\square$

4.3 Discrete gradient flow

For each discrete problem, we compute local minimisers using gradient descent, for which the basic result is the following (see [Bar15, §4.3.1]):

Theorem 4.10. (Projected gradient descent) *Let V_ε and J_ε^θ be given as in Problem 4.2 and let $(\cdot, \cdot)$ be the scalar product on V_ε. The map F_ε: $V_\varepsilon \times V_\varepsilon \to (V_\varepsilon \times V_\varepsilon)'$ given by*

$$F_\varepsilon^\theta[u_\varepsilon, z_\varepsilon](\varphi_\varepsilon, \psi_\varepsilon) := \theta \int_\omega Q_2^\varepsilon [\nabla_s u_\varepsilon + \tfrac{1}{2} z_\varepsilon \otimes z_\varepsilon, \nabla_s \varphi_\varepsilon + (z_\varepsilon \otimes \psi)_s] \, \mathrm{d}x$$

$$+ \frac{1}{12} \int_\omega Q_2[\nabla z_\varepsilon - B, \nabla \psi_\varepsilon] \, \mathrm{d}x$$

$$+ 2\mu_\varepsilon \int_\omega \operatorname{curl} z_\varepsilon \operatorname{curl} \psi_\varepsilon \, \mathrm{d}x, \tag{4.5}$$

is the Fréchet derivative of J_ε^θ. Let $\pi_u \colon V_\varepsilon^2 \to (V_\varepsilon \cap X_u)^2$ be the linear orthogonal projection onto its image. The sequence defined as

$$w_\varepsilon^{j+1} := w_\varepsilon^j + \alpha_j \pi_u d_\varepsilon^j,$$

with $w_\varepsilon^0 = (u_\varepsilon^0, v_\varepsilon^0) \in (V_\varepsilon \cap X_u)^2$ and $d_\varepsilon^j \in V_\varepsilon \times V_\varepsilon$ such that

$$(d_\varepsilon^j, \xi_\varepsilon) = -F_\varepsilon^\theta[w_\varepsilon^j](\xi_\varepsilon) \text{ for all } \xi_\varepsilon \in V_\varepsilon \times V_\varepsilon, \tag{4.6}$$

and α_j determined with line search is energy decreasing. A line search means computing the maximal $\alpha_j \in \{2^{-k} : k \in \mathbb{N}\}$ such that

$$J_\varepsilon^\theta(w_\varepsilon^j + \alpha_j \pi_u d_\varepsilon^j) \leqslant J_\varepsilon^\theta(w_\varepsilon^j) - \rho \alpha_j \|\pi_u d_\varepsilon^j\|_2^2,$$

where $\rho \in (0, 1/2)$ is the proverbial fudge factor.

Proof. For the computation of F_ε^θ see Appendix A.7, where we work with analogous functionals. To see that the iteration is energy decreasing use (4.6) and the self-adjointness of $\pi_u = \pi_u^2$ to compute

$$\frac{\mathrm{d}}{\mathrm{d}\alpha}\big|_{\alpha=0} J_\varepsilon^\theta(w_\varepsilon^j + \alpha \pi_u d_\varepsilon^j) = F_\varepsilon^\theta[w_\varepsilon^j](\pi_u d_\varepsilon^j) = -(\pi_u d_\varepsilon^j, \pi_u d_\varepsilon^j) \leqslant 0.$$

The existence of $\alpha_j > 0$ is guaranteed as long as $J_\varepsilon^\theta \in C^2(V_\varepsilon^2)$ because then we can perform a Taylor expansion and use again (4.6):

$$J_\varepsilon^\theta(w_\varepsilon^j + \alpha_j \pi_u d_\varepsilon^j) = J_\varepsilon^\theta(w_\varepsilon^j) - \alpha_j \|\pi_u d_\varepsilon^j\|_{\mathcal{S}}^2 + \mathcal{O}(\alpha_j^2). \qquad \Box$$

Remark 4.11. (Caveat: local and global minimisers) Even though we now know that the discrete energies correctly approximate the continuous one, as well as any global minimisers, gradient descent on each discrete problem is only guaranteed to converge to some local minimiser $w_\varepsilon^\star$. Lacking some means of tracking a particular $w_\varepsilon^\star$ as $\varepsilon \to 0$, there is not much one can do to prove that our method actually approximates the true global minimisers of $\mathcal{J}_{\mathrm{vK}}^\theta$. Unless $\theta \ll 1$, in which case we know all critical points to be global minimisers (cf. Theorem 3.12).[4.9]

4.4 Experimental results

For the implementation of the discretisation detailed above, we employ the FENICS library [ABH+15] in its version 2017.1.0. The code is available at [dB17b] and includes the model, parallel execution, experiment tracking using SACRED [GKC+17] with MONGODB as a backend and exploration of results with JUPYTER [KP+16] notebooks, OMNIBOARD [Sub18] and a custom application. Everything is packaged using DOCKER-COMPOSE for simple reproduction of the results and one-line deployment.

We set $\omega = \hat{B}_1(0)$, a (coarse) polygonal approximation of the unit disc and test several initial conditions. The space V_ε has ~7000 dofs. We implement a general Q_2 for isotropic homogeneous material with the two (scaled) Lamé constants set to those of steel at standard conditions. We apply neither body forces nor boundary conditions, but hold one interior cell to fix the value of the free constants. We compute minimisers for increasing values of θ and $\mu_\varepsilon \sim 1/\sqrt{\varepsilon}$ via projected gradient descent (onto the space of admissible functions $V_\varepsilon \cap X_u$) and examine the symmetry of the final solution. The choice $\varepsilon^{-1/2}$ has shown to provide the

---

4.9. One possibility to test the convergence of a sequence $(w_\varepsilon^\star)$ of *local* minimisers to a *local* minimiser of J^θ, would be to test whether there exists some fixed radius $\eta > 0$, independent of ε, such that each $w_\varepsilon^\star$ is a minimiser of J_ε^θ over the ball $B_\eta(w_\varepsilon^\star)$ (setting the functional to $+\infty$ outside it). $\|w_\varepsilon^\star\|$ being uniformly bounded, a subsequence would converge weakly to some $w_0^\star$ which would minimise J^θ locally.

fastest convergence results while keeping the violation of the constraint in the order of 10^{-4} (higher penalties have the expected effect of adversely affecting convergence). We track two magnitudes as measures of symmetry: on the one hand we compute the mean bending strain over the domain and on the other, as a second simple proxy we employ the quotient of the lengths of the principal axes.

The first initial configuration is the trivial deformation $y_\varepsilon^0 = 0$. Note that because the model is prestrained, the ground state is non-trivial and the plate "wants" to reach a lower energy state. In Figure 4.1 we depict the results of running the energy minimisation procedure for multiple values of θ.

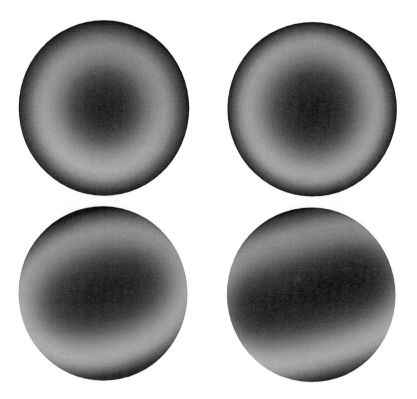

Fig. 4.1. Final configurations after gradient descent starting with a flat disk viewed from the top. From left to right, top to bottom: $\theta = 1, 81, 91$ and 150. Color represents the magnitude of the displacements $|w|$, from blue at its minimum to red at the maximum.

We further highlight the behaviour of the solution as a function of θ in Figures 4.2 and 4.3. In the first one we compute the mean bending strains

$$\frac{1}{|\omega|}\int_{\omega}(\nabla^2 v)_{ii}\,\mathrm{d}x \quad \text{with } i\in\{1,2\}.$$

As mentioned, these act as an easy to compute proxy for the (mean) principal curvatures. We observe how as θ increases both strains decrease almost by an equal amount as the body gradually opens up and flattens out, while retaining its radial symmetry. However, around $\theta\approx86$ a stark change takes place and one of the principal strains decreases while the other increases. This reflects the abrupt change of the minimiser to a cylindrical shape.

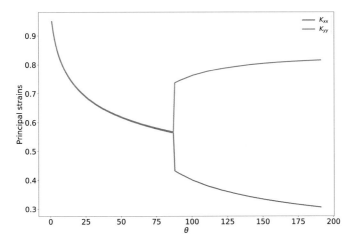

Fig. 4.2. Mean principal strains of the minimiser as a function of θ for the flat disk.

We observe the same phenomenon when we plot the quotient of the principal axes of the deformed disk in Figure 4.3.

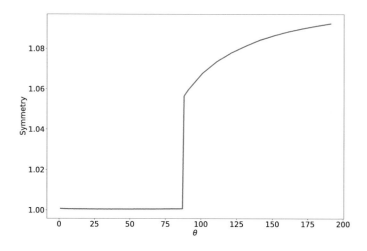

Fig. 4.3. Symmetry of the minimiser as a function of θ for the flat disk.

The second initial condition tested is an orthotropically skewed paraboloid. Basically, a spherical cap is pressed from the sides to obtain a "potato chip":

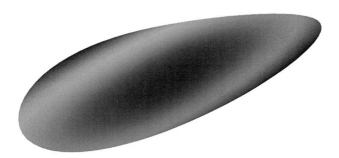

Fig. 4.4. A pringle in its initial configuration.

Testing this shape will highlight the effect of the initial configuration on the final curvature. We examine its symmetry in Figures 4.6 and 4.7

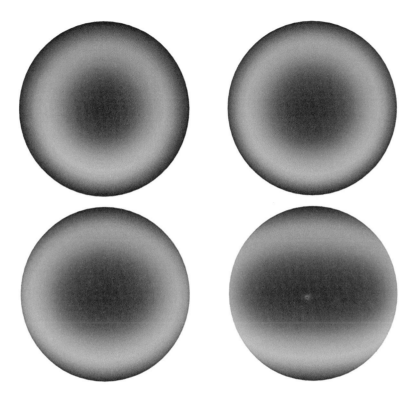

Fig. 4.5. Final states starting with skewed paraboloid. From left to right, top to bottom, $\theta = 1, 51, 61$ and 91.

Again there is a critical value of $\theta \approx 50$ around which the shape of the minimiser drastically changes. Note however how the change is now gradual and we see intermediate shapes.

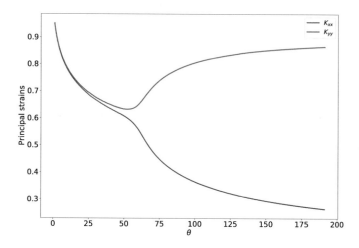

Fig. 4.6. Mean principal strains of the minimiser as a function of θ for the skewed paraboloid.

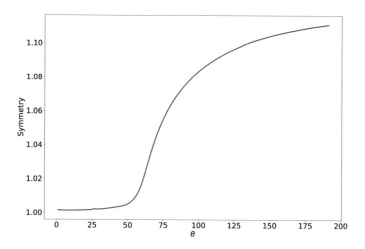

Fig. 4.7. Symmetry of the minimiser as a function of θ for the skewed paraboloid.

Appendix A
Auxiliary results

A.1 Some elementary matrix properties

We collect here standard definitions and properties of real matrices for the convenience of the reader. We refer to the monographs [HJ12, Bha09] for an extensive treatment of these ideas.

A.1.1 The norm of a real matrix

A function $|\cdot|: \mathbb{R}^{n \times n} \to \mathbb{R}$ is a **matrix norm** iff it is a positive, homogeneous function such that the triangle inequality and **submultiplicativity** hold, the latter meaning

$$|A\,B| \leqslant |A|\,|B|$$

for all matrices $A, B \in \mathbb{R}^{n \times n}$. Every matrix norm is consequently a vector norm, but the converse is not true, as the example $|A|_\infty := \max |A_{ij}|$ shows: consider $A_{ij} = 1$ with $n = 2$. Then $A^2 = 2\,A$ and this norm is not submultiplicative; however, $n|A|_\infty$ is.

The l^p vector norms for $p \in [1, \infty)$ do define matrix norms. In particular, the l^2 norm

$$|A| := \left(\sum_{i,j} A_{ij}^2\right)^{1/2}$$

is a matrix norm (submultiplicativity holds by Cauchy-Schwarz), which we call the **Frobenius norm**. It is induced by the **Hilbert-Schmidt inner product**

$$A : B := \operatorname{tr}(B^\top A),$$

since $A:A=\operatorname{tr}(A^{\mathsf{T}}A)=\sum_{i,j}A_{ij}^2$. That is:

$$|A|=\sqrt{\operatorname{tr}(A^{\mathsf{T}}A)}.$$

Remark A.1. Thisis not an **operator norm** in the sense that there is no vector norm $\|\cdot\|$ in $\mathbb{R}^n$ such that

$$\|\!|A|\!\|:=\sup_{x\neq0}\frac{\|Ax\|}{\|x\|}$$

is equal to $|A|$. This is because $|I|=\sqrt{n}>1$ for $n>1$ but a necessary condition for a matrix norm to be **induced** by a vector norm (i.e. to be an operator norm) is clearly $\frac{\|Ix\|}{\|x\|}\equiv1$.[A.1] Operator norms are automatically **compatible** with their inducing vector norms in the sense that $\|Ax\|\leqslant\|\!|A|\!\|\,\|x\|$. Despite its not being an operator norm, the Frobenius norm has this property since $|Ax|^2=\sum_{i,j}A_{ij}^2x_j^2\leqslant\sum_{i,j}A_{ij}^2|x^2|=|A|^2|x|^2$.

The l^2 or Frobenius norm is different from the norm *induced* by it, which can be proven to be the so called **spectral norm**: $|A|_2:=\max\{\sqrt{\lambda}:\lambda\in\operatorname{spectrum}(A^{\mathsf{T}}A)\}$.

Because of its definition, the following properties are useful when working with the Frobenius norm. For proofs, see e.g. [HJ12].

Proposition A.2. *Fundamental properties of the trace.*

1. $\operatorname{tr}(\alpha A+B)=\alpha\operatorname{tr}A+\operatorname{tr}B$.
2. $\operatorname{tr}A=\operatorname{tr}A^{\mathsf{T}}$.
3. $\operatorname{tr}(AB)=\operatorname{tr}(BA)$.
4. $\operatorname{tr}(ABC)=\operatorname{tr}(BCA)=\operatorname{tr}(CAB)$.
5. $\operatorname{tr}(S^{-1}AS)=\operatorname{tr}(A)$ *for every invertible matrix* $S\in\mathbb{R}^{n\times n}$.

Finally, we define the **absolute value** of $A\in\mathbb{R}^{n\times n}$ as the symmetric matrix:

$$[A]:=\sqrt{A^{\mathsf{T}}A}.$$

Lemma A.3. *(Some properties of the Frobenius norm).* *Let* $A,B\in\mathbb{R}^{n\times n}$:

1. $|A|=|A^{\mathsf{T}}|$.

A.1. Note that if we tried to fix this by scaling the Frobenius norm by $n^{-1/2}$, the property of submultiplicativity would be lost.

2. *It is invariant under orthogonal transformations. In particular it is invariant under rotations:* $|A| = |A\,U| = |U\,A|$ *for any orthogonal matrix* U.

3. *It is* **absolute**: $|A| = |[A]|$.

4. *It is* **monotone**: *if* $A \leqslant B$ *then* $|A| \leqslant |B|$.

5. $|A| = (\sum_{i=1}^{n} \mu_i^2)^{1/2}$, *where* μ_i *are the singular values of the matrix* A.

6. *If* $A = A^\top$, *then* $|A| = (\sum_{i=1}^{n} \lambda_i^2)^{1/2}$, *where* λ_i *are the eigenvalues of* A.

7. *If* A *is orthogonal* $(A^\top A = I)$, *then* $|A| = \sqrt{n}$.

We say that a symmetric matrix $A \in \mathbb{R}^{n \times n}$ is **positive semidefinite** if $x^\top A x \geqslant 0$ for all $x \neq 0$, and that it is **positive definite** if $x^\top A x > 0$ for all $x \neq 0$. A matrix A has a **square root** if there exists some matrix B such that $B^\top B = A$. For positive semidefinite matrices we can define a partial order

$$A \geqslant B \text{ iff } A - B \text{ is positive semidefinite.}$$

As we will see below, the Frobenius norm is monotone wrt. this order. Further properties are (for proofs see e.g. [Bha09]):

Proposition A.4. *Positive (semi-) definite matrices.*

1. *A matrix is positive definite, resp. semidefinite, iff all of its eigenvalues are positive, resp. non-negative.*

2. *A positive semidefinite matrix has at least one square root and exactly one which itself positive semidefinite.*

3. *A positive definite matrix is invertible and its inverse is itself invertible.*

Proposition A.5. *Some useful inequalities:*

1. $\text{tr}\, A \geqslant 0$ *for every positive semidefinite matrix* A.

2. $|\text{tr}\, A| \leqslant \text{tr}\, [\![A]\!]$.

3. $\text{tr}\,(A^\top B) \leqslant \frac{1}{2}(|A|^2 + |B|^2)$.

The **symmetric** and **antisymmetric part** of $A \in \mathbb{R}^{n \times n}$ are defined as

$$A_s = \frac{1}{2}(A + A^\top), \quad A_a = \frac{1}{2}(A - A^\top).$$

As a consequence of the definition one always has $A = A_s + A_a$ as well as

$$A_s \perp A_a$$

since $A_s\colon A_a = \operatorname{tr}(A_a^T A_s) = -\operatorname{tr}(A_a A_s) = -\operatorname{tr}(A_s A_a) = -A_a\colon A_s$. Therefore we have the following nice property (to prove it, compute $(A - A_s)\colon (A - A_s)$):

Lemma A.6. *Let* $A \in \mathbb{R}^{n \times n}$. *Then*

$$|A|^2 = |A_s|^2 + |A_a|^2.$$

A.1.2 Some matrix groups

We write $O(n)$ for the group of all orthogonal $n \times n$ matrices: $A^T A = I$. Its subgroup $SO(n)$ (the **special orthogonal group**) consists of those with positive determinant (i.e. the group of rotations in $\mathbb{R}^n$). By definition this set is bounded since $|A| = \sqrt{n}$ for all $A \in O(n)$. Finally $so(n)$ is the set of all real antisymmetric matrices $A = -A^T$.

The distance of a matrix $A \in \mathbb{R}^{n \times n}$ to any of the compact sets $\mathscr{G} \in \{O(n), SO(n), so(n)\}$ is given by

$$\operatorname{dist}(A, \mathscr{G}) = \min_{S \in \mathscr{G}} |A - S|.$$

From this definition and the triangle inequality it immediately follows that

$$\operatorname{dist}(A + B, \mathscr{G}) = \min_{S \in \mathscr{G}} |A + B - S| \leqslant |A| + \operatorname{dist}(B, \mathscr{G}). \tag{A.1}$$

Lemma A.7. *Let* $A \in \mathbb{R}^{n \times n}$. *Then*

$$\operatorname{dist}(A, so(n)) = |A_s|.$$

Proof. Let $S \in so(n)$. Because taking the symmetric and antisymmetric parts of a matrix are linear operations we have by Lemma A.6:

$$|A - S|^2 = |(A - S)_s|^2 + |(A - S)_a|^2 = |A_s|^2 + |A_a - S|^2.$$

But then

$$\operatorname{dist}(A, so(n)) = \min_{S \in so(n)} |A - S|^2 = |A_s|^2 + \min_{S \in so(n)} |A_a - S|^2 = |A_s|^2. \qquad \square$$

Proposition A.8. *Some properties of the distance to* $SO(n)$. *Let* $A \in \mathbb{R}^{n \times n}$:

1. For every rotation $R \in SO(n)$:

$$\operatorname{dist}(RA, SO(n)) = \operatorname{dist}(A, SO(n)). \tag{A.2}$$

2.

$$\operatorname{dist}(A, \mathrm{SO}(n)) \geqslant |[A] - I|. \tag{A.3}$$

3. *If* $\det A > 0$, *then equality holds:*

$$\operatorname{dist}(A, \mathrm{SO}(n)) = |[A] - I| = \operatorname{dist}([A], \mathrm{SO}(n)). \tag{A.4}$$

Proof. 1. If $R, S \in \mathrm{SO}(n)$ then by the invariance of the norm under orthogonal transformations, $|R\,A - S| = |R^{\mathsf{T}}\,(R\,A - S)| = |A - R^{\mathsf{T}}\,S|$. For any rotation R, one has $\mathrm{SO}(n) = \{R^{\mathsf{T}} S : S \in \mathrm{SO}(n)\}$, so that in fact

$$\operatorname{dist}(RA, \mathrm{SO}(n)) = \min_{S \in \mathrm{SO}(n)} |RA - S| = \min_{Q \in \mathrm{SO}(n)} |A - Q| = \operatorname{dist}(A, \mathrm{SO}(n)).$$

2. To prove (A.3) we let $S \in \mathrm{SO}(n)$ be arbitrary and compute using the properties of the trace:

$$
\begin{aligned}
|A - S|^2 &= \operatorname{tr}[(A - S)^{\mathsf{T}}(A - S)] \\
&= \operatorname{tr}[A^{\mathsf{T}} A + S^{\mathsf{T}} S - (S^{\mathsf{T}} A + A^{\mathsf{T}} S)] \\
&= \operatorname{tr}(A^{\mathsf{T}} A) + n - 2\operatorname{tr}(A^{\mathsf{T}} S). \\
&= \operatorname{tr}(A A^{\mathsf{T}}) + n - 2\operatorname{tr}(A S^{\mathsf{T}}).
\end{aligned}
$$

Therefore

$$\min_{S \in \mathrm{SO}(n)} |A - S|^2 = \operatorname{tr}(A A^{\mathsf{T}}) + n - 2 \max_{S \in \mathrm{SO}(n)} \operatorname{tr}(A S^{\mathsf{T}}).$$

Now using again the properties of the trace we have

$$\operatorname{tr}(A S^{\mathsf{T}}) \leqslant \operatorname{tr}([A S^{\mathsf{T}}]) = \operatorname{tr}([A]) \tag{A.5}$$

and consequently

$$\min_{S \in \mathrm{SO}(n)} |A - S|^2 \geqslant \operatorname{tr}(A A^{\mathsf{T}}) + n - 2\operatorname{tr}([A]) = \operatorname{tr}([A] - I)^2 = |[A] - I|^2.$$

3. Let $A = R\,[A]$ be the polar decomposition of A. Then $[A] = R^{\mathsf{T}} A$ and if $\det A > 0$ then we know that $R \in \mathrm{SO}(n)$ and the following inequality holds

$$\operatorname{tr}([A]) = \operatorname{tr}(R^{\mathsf{T}} A) = \operatorname{tr}(A R^{\mathsf{T}}) \leqslant \max_{S \in \mathrm{SO}(n)} \operatorname{tr}(A S^{\mathsf{T}}).$$

Consequently, using (A.5) we have

$$\operatorname{tr}([A]) = \max_{S \in \mathrm{SO}(n)} \operatorname{tr}(A S^{\mathsf{T}}),$$

and the first equality is proved. We can prove the second one with argu-ments analogous to the previous ones or we can use the polar decomposi-tion and (A.2):

$$\operatorname{dist}(A, \operatorname{SO}(n)) = \operatorname{dist}(R[A], \operatorname{SO}(n)) = \operatorname{dist}([A], \operatorname{SO}(n)). \qquad \square$$

A.1.2.1 A linearisation at the identity

Proposition A.9. *Linearising the function* $\operatorname{dist}(\cdot, \operatorname{SO}(n))$ *at the identity* I *yields:*

$$\operatorname{dist}(A, \operatorname{SO}(n)) = |A_s - I| + \mathcal{O}(|A - I|^2).$$

Proof. Note first that by the continuity of the determinant we may always find $\varepsilon > 0$ such that $|A - I| < \varepsilon$ implies $|\det A - \det I| < 1$, or $0 < \det A < 2$. Fix that ε.

Step 1: $|A - I| < \varepsilon$, *linearisation around* I.

The linearisation at 0 of the scalar function $x \mapsto \sqrt{1+x}$ is

$$\sqrt{1+x} - 1 = \frac{1}{2}x + \mathcal{O}(|x|^2).$$

Letting $A = I + G$ for some "small" G and proceeding by analogy with this equation we find

$$
\begin{aligned}
\sqrt{A^\mathsf{T} A} - I &= \sqrt{(I+G)^\mathsf{T}(I+G)} - I \\
&= \sqrt{I + (G^\mathsf{T}G + G^\mathsf{T} + G)} - I \\
&= \sqrt{I + X} - I \\
&= \frac{1}{2}(G^\mathsf{T}G + G^\mathsf{T} + G) + \mathcal{O}(|(G^\mathsf{T}G + G^\mathsf{T} + G)|^2) \\
&= \frac{1}{2}(G^\mathsf{T} + G) + \mathcal{O}(|G|^2) \\
&= \frac{1}{2}(A^\mathsf{T} + A) - I + \mathcal{O}(|A - I|^2).
\end{aligned}
$$

Because we have assumed that $\det A > 0$ we may apply Proposition A.8 above to conclude:

$$\operatorname{dist}(A, \operatorname{SO}(n)) = |A_s - I| + \mathcal{O}(|A - I|^2).$$

Step 2: $|A - I| > \varepsilon$.

We show that $|\mathrm{dist}\,(A, \mathrm{SO}(n)) - |A_s - I|| \leqslant C\,|A - I|^2$ for some $C > 0$ when $|A - I| > \varepsilon > 0$. The main idea is that the growth of $|A_s - I|$ is linear in $|A|$ and it is therefore possible to estimate it above by a multiple of $|A - I|^2$.

On the one hand, repeated application of the triangle inequality yields: $|A_s - I| \leqslant \frac{1}{2}(|A^\top| + |A|) + |I| = |A| + |I|$. On the other $|A_s - I| \geqslant \frac{1}{2}|A^\top + A| - |I| \geqslant \frac{1}{\sqrt{2}}|A| - |I|$, where the last lower bound is due to

$$\frac{1}{2}|A^\top + A| = \frac{1}{2}\sqrt{2\,\mathrm{tr}\,(A^2) + 2\,\mathrm{tr}\,(A^\top A)} \geqslant \frac{\sqrt{2}}{2}\sqrt{\mathrm{tr}\,(A^\top A)} = \frac{1}{\sqrt{2}}|A|.$$

By Proposition A.8 above and the definition of the distance we have $|[A] - I| \leqslant \mathrm{dist}\,(A, \mathrm{SO}(n)) \leqslant |A - I|$ and

$$\begin{aligned}
m_1 := |[A] - I| - (|A| + |I|) &\leqslant \mathrm{dist}\,(A, \mathrm{SO}(n)) - |A_s - I| \\
&\leqslant |A - I| - (2^{-1/2}|A| - |I|) =: m_2.
\end{aligned}$$

So $|\mathrm{dist}\,(A, \mathrm{SO}(n)) - |A_s - I|| \leqslant \max\{|m_1|, |m_2|\}$ and we want to show that this is of order $\mathcal{O}(|A - I|^2)$. More applications of the triangle inequality yield

$$|m_1| \leqslant |A| + |[A]| + 2\,|I| \overset{(a)}{=} 2\,(|A| + |I|) \overset{(b)}{\leqslant} C_1|A - I| \overset{(c)}{\leqslant} C_2|A - I|^2,$$

where:

a) We use that $|A| = |[A]|$.
b) We use that $|A - I| > \varepsilon > 0$:

$$|A| \leqslant |A - I| + |I| \leqslant \begin{cases} 2\,|A - I| & \text{if } |A - I| \geqslant |I|, \\ 2\,|I| = \dfrac{2\,|I|}{\varepsilon}\varepsilon \leqslant C|A - I| & \text{if } |A - I| < |I|. \end{cases}$$

c) We use again $|A - I| > \varepsilon$ and consider the two cases where $|A - I|$ is $\leqslant 1$ or > 1: If $\varepsilon^2 < |A - I|^2 < |A - I| < 1$ then $|A - I| < 1 < \frac{1}{\varepsilon}|A - I|^2 =: C|A - I|^2$. If $1 \leqslant |A - I|$, then $|A - I| \leqslant C|A - I|^2$ with $C = 1$.

And using the same method we obtain

$$|m_2| \leqslant 2\,|A| + 2\,|I| \overset{(b)}{\leqslant} C_1|A - I| \overset{(c)}{\leqslant} C_2|A - I|^2.$$

So indeed

$$|\mathrm{dist}\,(A, \mathrm{SO}(n)) - |A_s - I|| = \mathcal{O}(|A - I|^2). \qquad \square$$

A.1.2.2 The tangent space to SO(n)

Consider the map $f\colon \mathbb{R}^{n\times n} \to \mathbb{R}^{n\times n}$ defined as $f(A) = A^{\mathsf{T}} A - I$. With the trivial localisations $\phi\colon \mathbb{R}^{n\times n} \to \mathbb{R}^{n^2}$ we immediately see that f is differentiable because the components of $\phi \circ f \circ \phi^{-1}$ are simply polynomial functions of degree 2. Therefore the differential map at some point A and evaluated at B exists and coincides with the directional derivative at A along B:

$$
\begin{aligned}
\delta_B f(A) &:= \lim_{\lambda \to 0} \frac{f(A + \lambda B) - f(A)}{\lambda} \\
&= \lim_{\lambda \to 0} \frac{(A + \lambda B)^{\mathsf{T}}(A + \lambda B) - A^{\mathsf{T}} A}{\lambda} \\
&= \lim_{\lambda \to 0} \frac{A^{\mathsf{T}} A + \lambda^2 B^{\mathsf{T}} B + \lambda B^{\mathsf{T}} A + \lambda A^{\mathsf{T}} B - A^{\mathsf{T}} A}{\lambda} \\
&= B^{\mathsf{T}} A + A^{\mathsf{T}} B.
\end{aligned}
$$

In particular, the differential map at the identity is given by:

$$
Df[I](A) = A^{\mathsf{T}} + A.
$$

Obviously $f^{-1}\{0\} = O(n)$, the group of orthogonal matrices. Because of the implicit function theorem $O(n)$ is then a manifold, and it can be shown that it has dimension $n(n-1)/2$. The tangent space to $O(n)$ at I is consequently given by

$$
T_I O(n) = \ker Df[I] = \{ A \in \mathbb{R}^{n\times n} \colon A^{\mathsf{T}} + A = 0 \} = \mathrm{so}(n),
$$

the set of antisymmetric matrices. Because $\mathrm{SO}(n)$ is a subgroup of $O(n)$ we have $T_I \mathrm{SO}(n) = T_I O(n)$. At any other point $B \in \mathrm{SO}(n)$, the tangent space may be obtained as

$$
T_B \mathrm{SO}(n) = B \cdot \mathrm{so}(n) = \{ BA \colon A \in \mathrm{so}(n) \} = \{ AB \colon A \in \mathrm{so}(n) \} = \mathrm{so}(n) \cdot B.
$$

Another approach: Alternatively, we may consider vectors in the tangent space $T_I \mathrm{SO}(n)$ as derivatives $M = \gamma'(0)$ of smooth curves $\gamma\colon (-\delta, \delta) \to \mathrm{SO}(n)$ such that $\gamma(0) = I$. Then, because $\gamma(t) \in \mathrm{SO}(n)$ one has $\gamma(t)\gamma(t)^{\mathsf{T}} = I$ and differentiating $\gamma'(t)\gamma(t)^{\mathsf{T}} + \gamma(t)\gamma'(t)^{\mathsf{T}} = 0$. But then, for $t = 0$: $\gamma'(0) + \gamma'(t)^{\mathsf{T}} = 0$, i.e. $\gamma'(0) \in \mathrm{so}(n)$. This proves that $T_I \mathrm{SO}(n) \subset \mathrm{so}(n)$ and because both spaces have dimension $n(n-1)/2$ the inclusion is actually an equality.

The exponential map $A \mapsto e^A := \sum_k A^k/k!$ maps $so(n)$ onto $SO(n)$ (and it is surjective):

Lemma A.10. *Let $A \in so(n)$. Then e^A is a rotation.*

Proof. This follows from the properties of the exponential map (see e.g. [Gal01, Chapter 14]). For $A \in so(n)$ it holds $-A = A^\top$, therefore $(e^A)^\top = e^{A^\top} = e^{-A}$ and consequently $(e^A)^\top e^A = e^{-A+A} = I$ and $e^A (e^A)^\top = I$ as well, so A is orthogonal. Because $\det e^A = e^{\mathrm{tr}A}$ and $\mathrm{tr}\, A = 0$, it is a rotation. $\square$

A.2 On quadratic forms

A **quadratic form** is a polynomial function in several variables whose terms have all degree 2 (it is **homogeneous**). First we collect a few elementary properties of these objects, then prove several lemmas related to the elastic energy densities that we consider and their linearisations.

Proposition A.11. *Let $Q: \mathbb{R}^n \to \mathbb{R}$ be a quadratic form:*

1. *There exists a unique symmetric matrix $Q \in \mathbb{R}^{n\times n}_{\mathrm{sym}}$ such that $Q(x) = x^\top Q x$.*
2. *There exists a unique symmetric bilinear form $Q: \mathbb{R}^n \times \mathbb{R}^n \to \mathbb{R}$ such that $Q(x) = Q[x,x]$.*
3. *The following identity holds: $Q[x,y] = \frac{1}{2}(Q(x+y) - Q(x) - Q(y))$.*
4. *Q is positive semidefinite (i.e. $Q(x) \geqslant 0$ for all $x \neq 0$) iff it is convex.*
5. *If Q is positive semidefinite and not degenerate (i.e. $Q \not\equiv 0$), then it is positive definite over $\ker^\perp Q$.*
6. *There exists a constant $M > 0$ such that $Q(x) \leqslant M |x|^2$ for all $x \in \mathbb{R}^n$. If in addition Q is positive definite (i.e. $Q(x) > 0$ for all $x \neq 0$), then there exists a constant $m > 0$ such that $Q(x) \geqslant m |x|^2$ for all $x \in \mathbb{R}^n$.*

Proof. 1, 2, 3. Either of the first two statements may be taken as a definition for quadratic form over $\mathbb{R}^n$ and the third one is just a simple computation. Notice that we use the number of arguments (and the type of bracket) to distinguish between the quadratic form $Q(\cdot)$ and its associated bilinear form $Q[\cdot,\cdot]$.

4. This is immediate using the characterisation of convex functions with their Hessian: Fix any $x \in \mathbb{R}^n$. The first derivative is $DQ(x)[y] = 2\,Q[x, y]$ and the second one is the constant (as a function of x) bilinear map $D^2Q(x)[y, z] = Q[y, z] = y^\top \mathcal{Q} z$, where $\mathcal{Q}$ is the symmetric matrix associated to Q. If $\mathcal{Q}$ is positive semidefinite, so is $\mathcal{Q}$ and Q is convex. The reciprocal is then obvious.

5. This property follows after a coordinate transformation: because the associated $\mathcal{Q}$ is symmetric, there exists an orthogonal matrix P such that $\tilde{\mathcal{Q}} := P\,\mathcal{Q}\,P^\top = \mathrm{diag}(\lambda_1, \ldots, \lambda_n)$ with $0 \leqslant \lambda_i \in \sigma(\mathcal{Q})$ the spectrum of $\mathcal{Q}$. Define $\tilde{Q}$ to be the quadratic form associated to $\tilde{\mathcal{Q}}$. Then for any $\tilde{x} = P\,x$ in the new coordinates:

$$\tilde{Q}(\tilde{x}) = (Px)^\top (P\,\mathcal{Q}\,P^\top)(Px) = x^\top \mathcal{Q} x = Q(x).$$

Without loss of generality assume $\lambda_1, \ldots, \lambda_j$ to be zero and let λ_{j+1} be the smallest nonzero eigenvalue of $\mathcal{Q}$. In the corresponding basis of eigenvectors, the first j form a basis of $\mathcal{N}^\perp$. Therefore, for any $x \in \mathcal{N}^\perp$ the coordinates in this basis are $(0, \ldots, \tilde{x}_{j+1}, \ldots, \tilde{x}_n)$ and

$$Q(x) = \tilde{Q}(\tilde{x}) \geqslant \lambda_{j+1}\,(\tilde{x}_{j+1}^2 + \cdots + \tilde{x}_n^2) = \lambda_{j+1}\,|Px|^2 = \lambda_{j+1}\,|x|^2.$$

6. Define

$$M := \max_{|x|=1} Q(x) \quad \text{and} \quad m := \min_{|x|=1} Q(x).$$

And notice that for any $x \in \mathbb{R}^n$: $Q(x) = |x|^2 Q(x/|x|)$.   □

Lemma A.12. *Fix* $t \in \left(-\tfrac{1}{2}, \tfrac{1}{2}\right)$ *and let* $Q_3(t, \cdot)$ *be a positive semidefinite quadratic form in* $\mathbb{R}^{3\times 3}$. *Define*

$$Q_2(t, G) := \min_{c \in \mathbb{R}^3} Q_3(t, \hat{G} + c \otimes e_3), \;\text{ for all } G \in \mathbb{R}^{2\times 2},$$

where $\hat{G}$ *is the matrix in* $\mathbb{R}^{3\times 3}$ *given by* $\hat{G} := G_{\alpha\beta} e_\alpha \otimes e_\beta$ *for* $\alpha, \beta \in \{1, 2\}$ *and* $e_1, e_2 \in \mathbb{R}^3$ *are the first two standard basis vectors. Then* $Q_2(t, \cdot)$ *is a positive semidefinite quadratic form and there exists a linear map* $\mathscr{L}(t, \cdot): \mathbb{R}^{2\times 2} \to \mathbb{R}^3$ *such that*

$$Q_2(t, G) = Q_3(t, \hat{G} + \mathscr{L}(t, G) \otimes e_3)$$

for all $G \in \mathbb{R}^{2\times 2}$.

Proof. Fix some matrix $G \in \mathbb{R}^{2 \times 2}$. We omit the parameter t for brevity, but all the statements below apply pointwise. By convexity a critical point of Q_3 is a minimiser and the same is true of the restrictions to the affine subspaces

$$\{\hat{G} + c \otimes e_3 : c \in \mathbb{R}^3\} = \left\{ \begin{pmatrix} G_{11} & G_{12} & c_1 \\ G_{21} & G_{22} & c_2 \\ 0 & 0 & c_3 \end{pmatrix} : c_i \in \mathbb{R} \right\} \simeq \mathbb{R}^3.$$

Therefore, given any $G \in \mathbb{R}^{2 \times 2}$, finding the optimal $c^\star = c_{\min}(\hat{G})$ in the corresponding subspace is equivalent to solving for $c^\star$ in

$$\frac{\partial Q_3}{\partial c}(\hat{G} + c^\star \otimes e_3) = \vec{0}. \tag{A.6}$$

In fact, we prove that $c^\star = \mathscr{L}(G)$ for some linear function $\mathscr{L} \colon \mathbb{R}^{2 \times 2} \to \mathbb{R}^3$. Once this is established, we can define the linear map $\phi(G) := \hat{G} + \mathscr{L}(G) \otimes e_3$ and $Q_2[A, B] := Q_3[\phi(A), \phi(B)]$, where $Q_3[\cdot, \cdot]$ is the unique symmetric bilinear form associated with Q_3. Then $Q_2(G) = Q_3(\phi(G)) = Q_3[\phi(G), \phi(G)] = Q_2[G, G]$ and by linearity of ϕ, this is a bilinear map. Consequently $Q_2(G)$ is a quadratic form which is additionally positive semidefinite by construction: $Q_2(G) = \min Q_3(\hat{G} + c \otimes e_3) \geqslant 0$.

We need only show that $\mathscr{L}$ exists and is linear, but instead of computing the derivative in (A.6), we can use the directional derivatives at $\hat{G} + c \otimes e_3$ in direction $h \in \mathbb{R}^3$. Because

$$Q_3(\hat{G} + (c + h) \otimes e_3) = Q_3(\hat{G} + c \otimes e_3) + \underbrace{Q_3(h \otimes e_3)}_{= o(|h|^2)}$$

$$+ 2 Q_3[\hat{G} + c \otimes e_3, h \otimes e_3],$$

we have:

$$\frac{\partial Q_3}{\partial c}(\hat{G} + c \otimes e_3)[h] = 2 Q_3[\hat{G} + c \otimes e_3, h \otimes e_3] \text{ for every } h \in \mathbb{R}^3.$$

The condition that $c^\star$ fulfils is therefore

$$Q_3[\hat{G} + c^\star \otimes e_3, h \otimes e_3] = 0 \text{ for every } h \in \mathbb{R}^3.$$

Let now $G_1, \ldots, G_4 \in \mathbb{R}^{2 \times 2}$ be a basis. As argued above, to each one of the G_i corresponds by convexity a unique $c_i^\star$. Let now be $h \in \mathbb{R}^3$ and $G = \sum_i \alpha_i G_i, \alpha_i \in \mathbb{R}, i \in \{1, \ldots, 4\}$ be arbitrary. We check that $\mathscr{L}(G) = c^\star$

equals $\sum_i \alpha_i \mathscr{L}(G_i) = \sum_i \alpha_i c_i^\star$. From the previous equation we have on the one hand:

$$0 = Q_3[\hat{G}_i + c_i^\star \otimes e_3, h \otimes e_3] = \sum_i \alpha_i Q_3[\hat{G}_i + c_i^\star \otimes e_3, h \otimes e_3].$$

And on the other:

$$0 = Q_3[\sum_i \alpha_i \hat{G}_i + c^\star \otimes e_3, h \otimes e_3].$$

Equating both and cancelling the $\sum_i \alpha_i \hat{G}_i$ we obtain

$$0 = Q_3[\sum_i \alpha_i c_i^\star \otimes e_3 - c^\star \otimes e_3, h \otimes e_3].$$

Since h was arbitrary, this concludes the proof and we have a linear map $G \mapsto c^\star = \mathscr{L}(G)$. $\qquad\square$

Lemma A.13. *Fix* $t \in \left(-\frac{1}{2}, \frac{1}{2}\right)$. *Let* $Q_3(t, F) := D^2 W_0(t, I)[F, F]$ *with* W_0 *fulfilling the Assumptions 2.2 and define the relaxation* $Q_2(t, \cdot)$ *as in Lemma A.12:*

$$Q_2(t, G) := \min_{c \in \mathbb{R}^3} Q_3(t, \hat{G} + c \otimes e_3), \text{ for all } G \in \mathbb{R}^{2 \times 2},$$

where $\hat{G}$ *is the matrix in* $\mathbb{R}^{3 \times 3}$ *given by* $\hat{G} := G_{\alpha\beta} e_\alpha \otimes e_\beta$ *for* $\alpha, \beta \in \{1, 2\}$ *and* $e_1, e_2 \in \mathbb{R}^3$ *are the standard basis vectors. Then it holds that both* $Q_3(t, \cdot)$ *and* $Q_2(t, \cdot)$

1. are positive semidefinite (hence convex), and

2. vanish on antisymmetric matrices.

Proof.

1. Positive semidefiniteness of $Q_2(t, \cdot)$ follows directly from $Q_3(t, \cdot)$ having this property since $Q_2(t, G) = \min Q_3(t, \dots) \geqslant 0$. To prove that $Q_3(t, \cdot)$ is positive semidefinite simply notice that $Q_3(t, \cdot) = D^2 W_0(t, I)$, a positive semidefinite matrix because all rotations are minimisers of W_0 by Assumption 2.2.e.

2. Let $A = -A^\top \in \mathbb{R}^{n \times n}, t \in \left(-\frac{1}{2}, \frac{1}{2}\right)$ and $0 < \varepsilon \ll 1$. On the one hand:

$$
\begin{aligned}
W_0(t, I + \varepsilon A) &= \underbrace{W_0(t, I)}_{=0} + \underbrace{DW_0(t, I)[\varepsilon A]}_{=0} + \frac{1}{2}D^2 W_0(t, I)[\varepsilon A, \varepsilon A] \\
&\quad + o(\varepsilon^2) \\
&= \frac{\varepsilon^2}{2} Q_3(t, A) + o(\varepsilon^2).
\end{aligned}
$$

On the other hand, letting F be the projection of $I + \varepsilon A$ onto $SO(n)$ and $G = F - (I + \varepsilon A)$ one has, using (A.4) with $\det(I + \varepsilon A) > 0$ for $\varepsilon \ll 1$ and a linearisation of the square root (see the proof of Proposition A.9):

$$
\begin{aligned}
|G| &= \text{dist}(I + \varepsilon A, SO(n)) \\
&= \left| \sqrt{(I + \varepsilon A)^\top (I + \varepsilon A)} - I \right| \\
&= \left| \sqrt{I + \varepsilon^2 A^\top A} - I \right| \\
&= |I + \frac{1}{2}\varepsilon^2 A^\top A + o(\varepsilon^2) - I| \\
&= o(\varepsilon^2).
\end{aligned}
$$

Fig. A.1. The proof of Lemma A.13.b.

Therefore:

$$
W_0(t, I + \varepsilon A) = W_0(t, F - G) = \frac{1}{2}D^2 W_0(t, F)[G, G] + o(|G|^2) = o(\varepsilon^4),
$$

where we used that

$$
D^2 W_0(t, F)[G, G] = D^2 W_0(t, I)[GF^{-1}, GF^{-1}] = Q_3(t, GF^{-1}),
$$

and $|Q_3(t, GF^{-1})| \sim |GF^{-1}|^2 \sim |G|^2$. But then, by the previous result and dividing by ε^2:

$$
o(\varepsilon^4) = \frac{\varepsilon^2}{2} Q_3(t, A) + o(\varepsilon^2) \Rightarrow Q_3(t, A) = 0. \qquad \square
$$

Lemma A.14. *There exist constants $c_1, c_2 > 0$ such that for all $F \in \mathbb{R}^{3 \times 3}_{\text{sym}}$, $G \in \mathbb{R}^{2 \times 2}_{\text{sym}}$ it holds that*

$$Q_3(t, F) \geqslant c_1 |F|^2,$$

and

$$Q_2(t, G) \geqslant c_2 |G|^2,$$

uniformly in $t \in \left(-\frac{1}{2}, \frac{1}{2}\right)$.

Proof. Without loss of generality we can assume $|F| = 1$. Fix any $t \in \left(-\frac{1}{2}, \frac{1}{2}\right)$ and $1 \gg s > 0$ and use Assumptions 2.2.c and 2.2.e to compute

$$\begin{aligned}
\frac{1}{2} Q_3(t, sF) &\geqslant W_0(t, I + sF) - \omega(s) \\
&\geqslant c_1 \operatorname{dist}^2(I + sF, SO(3)) - \omega(s) \\
&= c_1 |sF|^2 - \omega(s).
\end{aligned}$$

Dividing both sides by s^2 and letting $s \to 0$ we obtain the first statement, while the second one holds by virtue of Q_2 being a minimal value of Q_3 over a subset of $\mathbb{R}^{3 \times 3}_{\text{sym}}$. ☐

Lemma A.15. *Let $m > 0$ and $B(m) := \{A \in \mathbb{R}^{2 \times 2}_{\text{sym}} : |A| \leqslant m\}$. Then*

$$|\mathscr{L}(t, A)| \leqslant C(m) \text{ for all } A \in B(m) \text{ uniformly in } t \in \left(-\frac{1}{2}, \frac{1}{2}\right).$$

Proof. This is an immediate consequence of the uniform bound of Lemma A.14. We have for all $|A| \in B(m)$:

$$Q_3(t, \hat{A} + \mathscr{L}(t, A) \otimes e_3) \gtrsim |\hat{A} + \mathscr{L}(t, A) \otimes e_3|^2 \gtrsim |\mathscr{L}(t, A) \otimes e_3|^2 - |\hat{A}|^2,$$

and the statement follows by Assumption 2.2.b. ☐

Lemma A.16. *Fix $t \in \left(-\frac{1}{2}, \frac{1}{2}\right)$. Let $Q_3 = Q_3(t, \cdot) := D^2 W_0(I)$ with W_0 fulfilling the Assumptions 2.2. The relaxation $Q_2 = Q_2(t, \cdot)$ from Lemma A.12 defines a scalar product in $L^2(\omega; \mathbb{R}^{2 \times 2}_{\text{sym}})$:*

$$\langle A, B \rangle := \int_\omega Q_2[A, B] \, dx.$$

Therefore for any $A, B \in L^2(\omega; \mathbb{R}^{2\times2}_{\text{sym}})$ and $\varepsilon > 0$ the following Cauchy inequalities hold:

$$\langle A,B \rangle \leqslant \frac{1}{2}\langle A,A \rangle + \frac{1}{2}\langle B,B \rangle, \tag{A.7}$$

$$\langle A,B \rangle \leqslant \varepsilon \langle A,A \rangle + \frac{1}{4\varepsilon}\langle B,B \rangle. \tag{A.8}$$

Additionally, we have Cauchy-Schwarz' inequality:

$$|\langle A,B \rangle| \leqslant \langle A,A \rangle^{1/2} \langle B,B \rangle^{1/2}. \tag{A.9}$$

Proof. $\langle \cdot, \cdot \rangle_{Q_2}$ is by construction bilinear and symmetric and it is positive definite on $L^2(\omega; \mathbb{R}^{2\times2}_{\text{sym}})$ by Lemma A.13. All inequalities are standard: for the first one use $0 \leqslant \langle A-B, A-B \rangle = \langle A,A \rangle + \langle B,B \rangle - 2\langle A,B \rangle$. For the second one write $\langle A, B \rangle = \langle (2\varepsilon)^{1/2} A, (2\varepsilon)^{-1/2} B \rangle \leqslant \frac{2\varepsilon}{2}\langle A, A \rangle + \frac{(2\varepsilon)^{-1}}{2}\langle B, B \rangle$. For the third one, define $\lambda := \langle A, B \rangle / \langle B, B \rangle$ and rearrange $0 \leqslant \langle A - \lambda B \rangle = \langle A,A \rangle + \lambda^2 \langle B,B \rangle - 2\lambda\langle A,B \rangle.$ $\square$

A.3 On geometric rigidity and Korn's inequality

In linear elasticity, the strain tensor is the symmetric gradient of the displacements, so that the energy is roughly $\int_\Omega |\nabla_s u|^2 \, \mathrm{d}x$. In order to apply the direct method of the calculus of variations to prove existence of minimisers, one requires bounds on infimizing sequences in order to extract (weakly) convergent subsequences, then some property (e.g. convexity) of the space of admissible functions is invoked to show that the limit remains in it. The bounds are obtained by means of the following:

Theorem A.17. (Korn's first inequality) *Let $\Omega \subset \mathbb{R}^n$ be an open bounded Lipschitz set, $n \geqslant 2$ and let $1 < p < \infty$. There exists a constant $C(p, \Omega)$ such that for all $w \in W^{1,p}(\Omega; \mathbb{R}^n)$*

$$\|\nabla w\|_{0,p} \leqslant C(p,\Omega)(\|\nabla_s w\|_{0,p} + \|w\|_{0,p}).$$

The second result of this form is of more interest to us, because it is the linear version of the estimate upon which all the proofs of compactness in the previous chapters rely.

Theorem A.18. (Korn's second inequality) *Let $\Omega \subset \mathbb{R}^n$ be an open bounded Lipschitz set, $n \geqslant 2$, and let $1 < p < \infty$. There exists a constant $C(p, \Omega)$ with the following property: For each $w \in W^{1,p}(\Omega; \mathbb{R}^n)$ there exists an anti-symmetric matrix $A_w \in so(n)$ such that*

$$\|\nabla w - A_w\|_{0,p} \leqslant C(p,\Omega) \, \|\nabla_s w\|_{0,p}.$$

Remark A.19. Korn's inequalities do not hold for $p = 1$ [Orn62].

Now, in Lemma A.7 it is shown that $\|\nabla_s w\|_{0,p} = \|\text{dist}(\nabla w, so(n))\|_{0,p}$, so the previous result tells us that we can bound $\|\nabla w\|$ by the distance to the space $so(n) := \mathbb{R}^{n \times n}_{skew}$ of the skew symmetric matrices after subtracting the right antisymmetric matrix. Recall that $so(n)$ is the tangent space to $SO(n)$ at the identity (see Section A.1.2.2), so the distance to $so(n)$ is a linearisation of the distance to $SO(n)$.

The nonlinear version by Friesecke, James and Müller [FJM02] of the previous theorem tells us that subtracting the right rotation one can bound $\|\nabla w\|$ by the distance to $SO(n)$. This has enabled the application of Γ-convergence methods to the derivation of rigorous lower dimensional theories. In an analogous role to its linear counterpart, the so-called *geometric rigidity estimate* implies the equicoerciveness of the scaled energies, i.e. precompactness of sequences of bounded energy.[A.2] The result is as follows:

Theorem A.20. ([FJM02, Theorem 3.1], Geometric rigidity) *Let $\Omega \subset \mathbb{R}^n$ be an open bounded Lipschitz set, $n \geqslant 2$. There exists a constant $C(\Omega)$ with the following property: for each $w \in W^{1,2}(\Omega; \mathbb{R}^n)$ there is an associated rotation $R_w \in SO(n)$ such that*

$$\|\nabla w - R_w\|_{0,2} \leqslant C(\Omega) \, \|\text{dist}(\nabla w, SO(n))\|_{0,2}.$$

A.2. We say that a family of functionals $F^h \colon X \to \overline{\mathbb{R}}$ is **equicoercive** if for all $c \in \mathbb{R}$ there exists a compact set $K_c \subset X$ such that $\{F^h \leqslant c\} \subset K_c$. Or, equivalently (in 1st countable topologies), if every sequence (x^h) with $F^h(x^h) \leqslant c$ has a convergent subsequence.

In the same vein that this a nonlinear version of Korn's second inequality, the following corollary plays the same role wrt. Korn's first inequality:

Corollary A.21. (**[JS13, Theorem A.8]**) *Let $\Omega \subset \mathbb{R}^n$ be an open bounded Lipschitz set, $n \geqslant 2$ and let $1 < p < \infty$. There exists a constant $C(p, \Omega)$ such that for all $w \in W^{1,p}(\Omega; \mathbb{R}^n)$*

$$\|\nabla w\|_{0,p} \leqslant C \left(\|\operatorname{dist}(I + \nabla w, \operatorname{SO}(n))\|_{0,p} + \|w\|_{0,p} \right).$$

Both estimates hold for $p \in (1, \infty)$ [CS06, p.854], but as in the linear case, not for $p = 1$ [CFM05].

We conclude this section with the following standard Korn-Poincaré estimate, which is an essential ingredient in the proof of Theorem 3.10:

Corollary A.22. *Let*

$$X_u := \left\{ u \in W^{1,2}(\omega; \mathbb{R}^2) : \int_\omega \nabla_a u = 0 \text{ and } \int_\omega u = 0 \right\}.$$

There is a constant $C = C(\omega) > 0$ such that

$$\|u\|_{0,2} \leqslant C \|\nabla_s u\|_{0,2} \text{ for all } u \in X_u.$$

Consequently, there is another constant $C = C(\omega) > 0$ such that

$$\|u\|_{1,2} \leqslant C \|\nabla_s u\|_{1,2} \text{ for all } u \in X_u.$$

Proof. Notice that the second statement follows immediately from Poincaré's inequality (recall that $u \in X_u$ implies $\int u = 0$), Korn's first inequality (Theorem A.17) and the first statement:

$$\|u\|_{1,2}^2 \lesssim \|\nabla u\|_{0,2}^2 \lesssim \|\nabla_s u\|_{0,2}^2 + \|u\|_{0,2}^2 \lesssim \|\nabla_s u\|_{0,2}^2.$$

We now prove the first estimate. For this we may assume $\|u\|_{0,2} = 1$ since for general u one simply applies the result to $u / \|u\|_{0,2}$. Suppose then that the inequality does not hold, that is: we can find scalars $C_n \to 0$ and functions $u_n \in X_u$ such that $\|u_n\|_{0,2} = 1$ and $\|\nabla_s u_n\|_{0,2} \leqslant C_n$. By Korn's first inequality the u_n are bounded in $W^{1,2}$:

$$\|u_n\|_{1,2} \lesssim \|\nabla_s u_n\|_{0,2} + \|u_n\|_{0,2} \leqslant C,$$

hence a subsequence exists (not relabelled), which converges weakly in $W^{1,2}$ to some u_0. But the constraints being linear the set X_u is convex, thus $u_0 \in X_u$. Now, the map $u \mapsto \|\nabla_s u\|_{0,2}^2$ is w.s.l.s.c.[A.3] and we have with Korn's second inequality (Theorem A.18) that

$$C \min_{A \in so(2)} \|\nabla u_0 - A\|_{0,2}^2 \leqslant \|\nabla_s u_0\|_{0,2}^2 \leqslant \liminf_{n \to \infty} \|\nabla_s u_n\|_{0,2}^2 = 0.$$

Therefore $\nabla u_0 = A \in so(2)$ and $u_0(x) = A x + b$. But, because $u_0 \in X_u$ it holds that $0 = \int_\omega \nabla_a u = \int_\omega A$ so A must be zero, and $0 = \int_\omega u$ so b must be zero as well, i.e. $u_0 = 0$.

Finally, by the compact embedding $W^{1,2}(\omega) \hookrightarrow L^2(\omega)$ [AF03, §6.3], a subsequence of (u_n) converges strongly in L^2 to u_0, a contradiction to the assumption $\|u_n\|_{0,2} = 1$. $\qquad\square$

A.4 Convergence boundedly in measure

Definition A.23. *Let Ω be a σ-finite measure space and let (f_n) be a sequence in $L^\infty(\Omega)$ converging in measure to $f \in L^\infty(\Omega)$. We say that f_n converges* **boundedly in measure** *provided that $f_n \to f$ in measure and* $\sup \|f_n\|_\infty < \infty$.

Here is the application which we require:

Lemma A.24. *Let $\Omega \subset \mathbb{R}^d$ be open and bounded, $f, f_n \in L^2(\Omega)$ and $f_n \to f$ boundedly in measure. Let $G, G_n \in L^2(\Omega)$ too and $G_n \rightharpoonup G$ weakly in $L^2(\Omega)$. Then $f_n G_n, fG \in L^2(\Omega)$ and*

$$f_n G_n \rightharpoonup fG \text{ in } L^2(\Omega).$$

A.3. Indeed, it is clearly convex and continuous, so its epigraph is convex and closed, consequently weakly closed, and this happens iff the function is w.s.l.s.c.

Proof. Let $M := \sup_n \|f_n\|_\infty$. Observe first that $\int_\Omega |f_n\, G_n|^2\, dx \leqslant M^2\, \|G_n\|_2^2 < \infty$ and, analogously, the product $fG \in L^2$. Therefore these are elements of the dual $(L^2)' \simeq L^2$ and we may indeed attempt to show weak convergence. To this end we take $\psi \in L^2(\Omega)$ arbitrarily and estimate the integral

$$\int_\Omega (G_n f_n \psi - Gf\,\psi)\, dx = \underbrace{\int_\Omega (G_n f_n \psi - G_n f\,\psi)\, dx}_{=:A_n}$$

$$+ \underbrace{\int_\Omega (G_n f\,\psi - Gf\,\psi)\, dx.}_{=:B_n}$$

On the one hand, $f\,\psi \in L^2$ thanks to f being uniformly bounded, and the L^2 weak convergence $G_n \rightharpoonup G$ then yields $B_n = \int_\Omega (G_n - G) f\,\psi\, dx \to 0$. On the other hand, we use Hölder's inequality and the fact that weakly convergent sequences are bounded: $A_n \leqslant \|G_n\|_2\, \|f_n\,\psi - f\,\psi\|_2 \leqslant C\, \|f_n\,\psi - f\,\psi\|_2$.

To estimate the integral $\|f_n\,\psi - f\,\psi\|_2$ we note that the sequence $f_n\,\psi \to f\,\psi$ in measure because $\psi \in L^2$ and the product of sequences converging in measure (in a finite measure space) converges in measure as well. Therefore $(f_n\,\psi - f\,\psi)^2 \to 0$ in measure too and this last sequence is dominated by the function $M\, \|\psi\|_{L^2}^2$, which is integrable because $|\Omega| < \infty$. By the dominated convergence theorem:

$$\int_\Omega (f_n\,\psi - f\,\psi)^2\, dx \to 0,$$

and the proof is complete. $\qquad\qquad\qquad\qquad\qquad\qquad\qquad\qquad\square$

A.5 Γ-convergence via maps

In Definition 2.5 we employed the technical device of P^h maps to encode the estimate required for compactness and the lower bounds of Section 2.3. In this section we gather some results and remarks on Γ-convergence in that context, while repeating some standard arguments where required. We consider proper functionals

$$F^h: Y \to \overline{\mathbb{R}} \text{ and } F: X \to \overline{\mathbb{R}},$$

such that $(F^h)_{h>0}$ Γ-converges to F via maps P^h as in Definition 2.5. We say that a functional $F: X \to \overline{\mathbb{R}}$ is **proper** if $F \not\equiv +\infty$ and $F(w) > -\infty$ always.

Remark A.25. For the computation of the lower bound, we may suppose that the sequence $(y^h)_{h>0}$ P^h-converging to $w \in X$ has finite energy, i.e. that $F^h(y^h)$ is uniformly bounded. Indeed, by passing to a subsequence $(y^{h'}) \subset (y^h)$, we have

$$\lim_{h' \to 0} F^{h'}(y^{h'}) = \liminf_{h \to 0} F^h(y^h),$$

and since we want to show that the right hand side is greater or equal than a finite value (the limit energy at w) we may assume that there exists some $C > 0$ such that $C > F^{h'}(y^{h'})$ or the estimate is trivial.

Remark A.26. Again for the computation of the lower bound, it is enough to show that for every P^h-convergent sequence $(y^h)_{h>0}$, there exists a subsequence $(y^{h'})_{h'>0}$ P^h-converging to the same limit with $\liminf_{h \to 0} F^{h'}(y^{h'}) = \lim_{h \to 0} F^{h'}(y^{h'})$ and such that the lower bound holds. To see why, assume one has this but the lower bound does not hold to arrive at a contradiction after extracting a minimising sequence. This implies that it is not a problem to take subsequences as needed along the way as we do in all the proofs.

The following statement turns out to be very useful for all upper bounds [Bra06, Remark 2.8].

Lemma A.27. *It is enough to define the recovery sequences on a subset $\mathscr{A}_0 \subset X_0$ which is dense wrt. the topology in which the limit functional is continuous.*

Recall that the three dimensional problems in Ω_h may have non-unique minimisers or none at all. Therefore we must consider sequences of **almost minimisers**, i.e. sequences $(y^h)_{h>0} \subset Y$ such that:

$$\limsup_{h \to 0} \left(F^h(y^h) - \inf_Y F^h \right) = 0.$$

The standard argument, which can be found in any textbook, for convergence of the almost minimal energies now goes as follows:

1. Boundedness: almost minimising sequences have bounded energy (Lemma A.28).
2. Compactness: sequences of bounded energy have P^h-convergent subsequences (Lemma A.30).
3. Convergence: P^h-convergent, almost minimising (sub)sequences have energies converging to the limit energy and their limit is a minimiser of the limit energy (Lemma A.29).

Lemma A.28. *(Boundedness) Any sequence of almost minimisers of F^h has bounded energy.*

Proof. Let $(y^h)_{h>0}$ be a sequence of almost minimisers:

$$\operatorname*{lsup}_{h\to 0}\left(F^h(y^h) - \inf_X F^h\right) = 0 \Leftrightarrow F^h(y^h) = \inf_X F^h + o(1).$$

Because the limit energy is a proper functional (i.e. $\not\equiv \infty$) there exists $w \in X$ with $F(w) < \infty$. Now let $(w^h)_{h>0}$ be a recovery sequence for w, i.e. $F^h(w^h) = F(w) + o(1)$ as $h \to 0$, then:

$$F^h(y^h) = \inf_X F^h + o(1) \leqslant F^h(w^h) + o(1) = F(w) + o(1) \leqslant C. \qquad \square$$

Lemma A.29. *(Fundamental theorem) Let $(y^h)_{h>0}$ be a sequence of almost minimisers such that $P^h(y^h) \to w$ in X as $h \to 0$. Then*

$$F^h(y^h) \to F(w) \text{ and } F(w) = \min_X F.$$

Proof. Let $\hat{w} \in X$ be arbitrary. By Γ-convergence there exists a recovery sequence $(\hat{y}^h)_{h>0}$. Then:

$$F(\hat{w}) = \lim_{h\to 0} F^h(\hat{y}^h) \geqslant \operatorname*{lsup}_{h\to 0}\inf_X F^h = \operatorname*{lsup}_{h\to 0} F^h(y^h) \geqslant \operatorname*{linf}_{h\to 0} F^h(y^h) \geqslant F(w),$$

which means that w minimises F. Taking now $\hat{w} = w$ we see that we must have equality everywhere, in particular in the last two steps, and consequently $\lim_{h\to 0} F^h(y^h) = F(w)$. $\qquad \square$

Notice that we can apply the preceding Lemma in the situation of Chapter 2 and in particular Theorem 2.6.

A.6 Compactness and identification of the limit strain

We collect here some known results proving compactness of sequences of scaled energy and providing explicit representations for the limit strains, as required for the proofs of Γ-convergence in Chapter 2. We recall the definition of the scaled elastic energies (2.1):

$$\mathscr{I}_\alpha^h(y) = \frac{1}{h^{2\alpha-2}} \int_{\Omega_1} W_0(x_3, \nabla_h y(x)(I + h^{\alpha-1} B^h(x_3))) \, dx.$$

The first Lemma shows that there are P^h-converging sequences (Definition 2.4):

Lemma A.30. ([FJM06, Lemma 1]) *Let $\alpha \in (2, \infty)$ and let $(y^h)_{h>0} \subset Y$ have finite scaled $\mathscr{I}_\alpha^h$ energy. For every $h > 0$ there exist constants $\overline{R}^h \in$ SO(3) and $c^h \in \mathbb{R}^3$ such for the corrected deformations*

$$\tilde{y}^h = \rho(y^h) := (\overline{R}^h)^\top y^h - c^h. \tag{A.10}$$

there exist rotations $R^h \colon \omega \to \mathrm{SO}(3)$ (extended constantly along x_3 to all of Ω_1 outside $\{0\} \times \omega$) approximating $\nabla_h \tilde{y}^h$ in $L^2(\Omega_1)$. Quantitatively:

$$\|\nabla_h \tilde{y}^h - R^h\|_{0,2,\Omega_1} \leqslant C h^{\alpha-1}.$$

By the invariance of the norm by rotations this implies

$$(R^h)^\top \nabla_h \tilde{y}^h \to I \text{ in } L^2(\Omega_1) \text{ as } h \to 0.$$

Furthermore,

$$\|R^h - I\|_{0,2,\Omega_1} \leqslant C h^{\alpha-2}.$$

Finally there exists a subsequence (not relabelled) such that for the scaled and averaged in-plane and out-of-plane displacements from (2.7) there exist $(u, v) \in W^{1,2}(\omega; \mathbb{R}^2) \times W^{2,2}(\omega)$ such that, if $\alpha \neq 3$:

$$u_\alpha^h \rightharpoonup u \text{ in } W^{1,2}(\omega; \mathbb{R}^2) \quad \text{and} \quad v_\alpha^h \to v \text{ in } W^{1,2}(\omega),$$

If $\alpha = 3$ an analogous result holds with u_θ^h and v_θ^h from (2.8).

Proof. This is exactly a particular case of [FJM06, Lemma 1], estimates (84) and (85) and estimates (86) and (87), once we prove that if $(y^h)_{h>0}$ have finite scaled $\mathscr{I}_\alpha^h$ energy, then they have finite scaled energy in the sense of [FJM06].

Note first that among all choices we can make for the energy density W which fulfil the assumptions in [FJM06], we can pick $\mathrm{dist}^2(\cdot, \mathrm{SO}(3))$. Therefore we will bound this quantity. Write $d(F) := \mathrm{dist}(F, \mathrm{SO}(3))$. We begin by using Assumption 2.2.e:

$$Ch^{2\alpha-2} \geqslant \int_{\Omega_1} W_0(x_3, \nabla_h y(x)(I + h^{\alpha-1} B^h(x_3)))$$

$$\gtrsim \int_{\Omega_1} d^2(\nabla_h y(x)(I + h^{\alpha-1} B^h(x_3))).$$

Consider now the following:

$$d^2(F(I + h^{\alpha-1} B^h)) \geqslant \tfrac{1}{2} d^2(F) - |F h^{\alpha-1} B^h|^2$$

$$\geqslant \tfrac{1}{2} d^2(F) - C h^{2\alpha-2} |1 + d^2(F)|$$

$$\geqslant \tfrac{1}{4} d^2(F) - C h^{2\alpha-2}.$$

But then we are done since:

$$h^{2\alpha-2} \gtrsim \int_{\Omega_1} \tfrac{1}{4} d^2(\nabla_h y). \qquad \square$$

Lemma A.31. ([FJM06, Lemma 2]) [A.4]*Let $\alpha \in (2, \infty)$ and let $(y^h)_{h>0}$ be a sequence in Y which P^h-converges to $(u, v) \in X_\alpha$ in the sense of Definition 2.4 with $R^h \colon \omega \to \mathrm{SO}(3)$ (extended constantly along x_3 to all of Ω_1 outside $\{0\} \times \omega$) such that*

$$\|\nabla_h y^h - R^h\|_{0,2,\Omega_1} \leqslant C h^{\alpha-1}. \qquad (\star)$$

Then:

$$A^h := \frac{1}{h^{\alpha-2}}(R^h - I) \longrightarrow \begin{cases} \sqrt{\theta}\, A & \text{if } \alpha = 3, \\ A & \text{else}, \end{cases} \quad \text{in } L^2(\omega; \mathbb{R}^{3\times 3}), \qquad (A.11)$$

where

$$A := e_3 \otimes \hat{\nabla} v - \hat{\nabla} v \otimes e_3,$$

and

$$G^h := \frac{(R^h)^\top \nabla_h y^h - I}{h^{\alpha-1}} \rightharpoonup G \text{ in } L^2(\Omega_1; \mathbb{R}^{3\times 3}), \qquad (A.12)$$

A.4. This is almost **word for word** [FJM06, Lemma 2] with the very minor addition of the factors θ, $\sqrt{\theta}$. For other scaling choices see [FJM06, p.208]. Note that this result is inspired by [Cia97, Theorem 5.4.2], which is itself based on [Cia97, Theorem 1.4.1.c].

where the submatrix $\check{G} \in \mathbb{R}^{2\times2}$ *is affine in* x_3:

$$\check{G}(x',x_3) = G_0(x') + x_3\,G_1(x') \tag{A.13}$$

and

$$G_1 = \begin{cases} -\sqrt{\theta}\,\nabla^2 v & \text{if } \alpha = 3, \\ -\nabla^2 v & \text{else,} \end{cases} \tag{A.14}$$

$$\operatorname{sym} G_0 = \begin{cases} \theta\left(\nabla_s u + \frac{1}{2}\nabla v \otimes \nabla v\right) & \text{if } \alpha = 3, \\ \nabla_s u & \text{if } \alpha > 3, \end{cases} \tag{A.15}$$

and

$$\nabla_s u + \frac{1}{2}\nabla v \otimes \nabla v = 0, \text{ if } \alpha \in (2,3).$$

Proof. See [FJM06, p.208 - 209].   □

A.7 Derivatives galore

Let $F\colon X \to Y$ be a map between Banach spaces and $U \subset X$ open. We say that F **is Gâteaux differentiable at** $w \in U$ if the limit

$$\delta F(w;\eta) := \lim_{\varepsilon \to 0} \frac{F(w+\varepsilon\,\eta) - F(w)}{\varepsilon} = \frac{\mathrm{d}}{\mathrm{d}\varepsilon}\Big|_{\varepsilon=0} F(w+\varepsilon\,\eta)$$

exists for all $\eta \in X$ and the map $\eta \mapsto \delta F(w;\eta)$ is linear and bounded from X to Y. We say that F **is Gâteaux differentiable in** U iff for every $w \in U$ the map $\delta F(w) \in \mathscr{L}(X,Y)$. We say that F **is continuously Gâteaux differentiable in** U if the map $w \mapsto \delta F(w)$ from U into $\mathscr{L}(X,Y)$ is continuous.

The following result relates the directional derivative to the differential map in the sense of Fréchet. For a proof see e.g. [Wer07, Satz III.5.4].

Lemma A.32. *Let* $F\colon X \to Y$ *be continuously Gâteaux differentiable in an open set* $U \subset X$. *Then* F *is Fréchet differentiable and its Fréchet differential at any* $w \in U$ *is* $DF(w) = \delta F(w)$.

Given some fixed direction, the first Gâteaux derivative or **first variation** is just a directional derivative, but for higher orders it is defined as

$$\delta^n F(w;\eta) := \frac{\mathrm{d}^n}{\mathrm{d}\varepsilon^n}\Big|_{\varepsilon=0} F(w+\varepsilon\,\eta),$$

whereas the iterated directional derivative is, e.g. for order 2:

$$\begin{aligned}
d^2F(w;\eta_1,\eta_2) &= \frac{d^2}{d\varepsilon_1\,d\varepsilon_2}\Big|_{\varepsilon_i=0} F(w+\varepsilon_1\eta_1+\varepsilon_2\eta_2)\\
&= \delta(\delta F(w;\eta_1))(w;\eta_2)\\
&= \lim_{\varepsilon\to 0} \frac{\delta F(w+\varepsilon\,\eta_2;\eta_1)-\delta F(w;\eta_1)}{\varepsilon}.
\end{aligned}$$

If the limits exist, this defines for each $w \in U$ a map $d^2F(w) = \delta(\delta(w))$: $X \times X \to Y$, which obviously cannot be equal to the map $\delta^2 F(w;\eta): X \to Y$. We have however the following:

Lemma A.33. *Let $F: X \to Y$ and $U \subset X$ open and assume that for each $w \in U$, the map $d^2F(w): X \times X \to Y$ is a continuous bilinear form. If in addition the map $w \mapsto d^2F(w)$ is continuous from U to $\mathscr{L}(X \times X, Y)$, then F is twice Fréchet differentiable and its second Fréchet differential at any $w \in U$ is $D^2F(w) = d^2F(w)$.*

Proof. Since $w \mapsto d^2F(w) = \delta(\delta F)(w) = d(\delta F)(w)$ is assumed to be continuous, by Lemma A.32, δF is Fréchet differentiable and $d(\delta F) = D(\delta F)$ on U. In particular, $\delta F = dF$ is also continuous over U, hence $\delta F = DF$, again by Lemma A.32. Consequently $D^2F(w) = d^2F(w)$ for all $w \in U$. $\square$

Finally, we state the classical Implicit Function Theorem, which plays a fundamental role in Section 3.2.1. For a proof see e.g. [Lan69, p. 125].

Theorem A.34. (Implicit Function Theorem) *Let X, Y, Z be Banach spaces and let $W \subset X, \Theta \subset Z$ be open sets. Let*

$$f: W \times \Theta \to Y$$

be a C^p map for $p \geqslant 1$ and assume there is a point $(w_0,\theta_0) \in W \times \Theta$ such that the partial derivative

$$\partial_w f(w_0,\theta_0): W \to Y$$

is an isomorphism and $f(w_0,\theta_0) = 0$. Then there exists a continuous map $\phi: \Theta_0 \to W$ defined on an open neighbourhood Θ_0 of θ_0 such that $\phi(\theta_0) = w_0$ and such that

$$f(\phi(\theta),\theta) = 0$$

for all $\theta \in \Theta_0$. If Θ_0 is sufficiently small, then ϕ is uniquely determined, and is also of class C^p.

A.7.1 A few computations

Without much attention to technicalities (e.g. differentiation under the integral sign), in this section we compute the first and second variation and Fréchet derivatives of the limit functional

$$I(w) := \frac{\theta}{2} \int_\omega \underbrace{Q_2(\nabla_s u + \frac{1}{2} \nabla v \otimes \nabla v)}_{A} + \frac{1}{24} \int_\omega Q_2(\nabla^2 v - I),$$

which was derived in (2.13) using the *t-independent* energy

$$W(t, F) := W_0\big(F\big(I + h^2 \sqrt{\theta}\, t I\big)\big),$$

with $\theta > 0$ and the misfit

$$B^h(t) = t \operatorname{Id}_3.$$

Throughout we assume $w \in X := W^{1,2}(\omega; \mathbb{R}^2) \times W^{2,2}(\omega)$.

With the notation of Lemma A.16, $\langle A, B \rangle := \int_\omega Q_2[A, B]$ and $\langle A \rangle := \langle A, A \rangle$, we collect some results in preparation for the computations to follow:[A.5]

$$\frac{\mathrm{d}}{\mathrm{d}\varepsilon}\big|_{\varepsilon=0} \langle A + \varepsilon B \rangle = 2 \langle A + \varepsilon B, B \rangle\big|_{\varepsilon=0} = 2\langle A, B \rangle,$$

$$\frac{\mathrm{d}}{\mathrm{d}\varepsilon}\big|_{\varepsilon=0} \langle A + \varepsilon B + \varepsilon^2 C \rangle = 2 \langle A + \varepsilon B + \varepsilon^2 C, B + 2\varepsilon C \rangle\big|_{\varepsilon=0}$$
$$= 2\langle A, B \rangle,$$

and

$$\frac{\mathrm{d}^2}{\mathrm{d}\varepsilon^2}\big|_{\varepsilon=0} \langle A + \varepsilon B \rangle = 2 \langle B \rangle,$$

A.5. Recall that if $Q[\cdot, \cdot]$ is a bilinear form in $\mathbb{R}^{n \times n}$ and $F, G \in \mathbb{R}^{n \times n}$ are functions of ε, then

$$\partial_\varepsilon Q[F, G] = Q[\partial_\varepsilon F, G] + Q[F, \partial_\varepsilon G]$$

as can be readily seen differentiating the expression in coordinates: $Q[F, G] = F_{ij} Q_{ijkl} G_{kl}$. If $Q_2(\cdot)$ is now a quadratic form with associated symmetric bilinear form $Q_2[\cdot, \cdot]$ one has

$$\partial_\varepsilon Q_2(F) = 2 Q_2[\partial_\varepsilon F, F] \quad \text{and} \quad \partial^2_{\varepsilon\varepsilon} Q_2(F) = 2 Q_2[\partial^2_{\varepsilon\varepsilon} F, F] + 2 Q_2(\partial_\varepsilon F).$$

$$\frac{\mathrm{d}^2}{\mathrm{d}\varepsilon}\big|_{\varepsilon=0}\langle A+\varepsilon B+\varepsilon^2 C\rangle = 2\langle 2C, A+\varepsilon B+\varepsilon^2 C\rangle\big|_{\varepsilon=0}$$
$$+2\langle B+2\varepsilon C\rangle\big|_{\varepsilon=0}$$
$$= 4\langle A,C\rangle+2\langle B\rangle.$$

First variation: we begin with the first variation of I along $\eta = (\varphi, \psi) \in W^{1,2}(\omega; \mathbb{R}^2) \times W^{2,2}(\omega)$, applying where necessary the Sobolev embeddings $W^{2,2} \hookrightarrow W^{1,4}$ and $W^{1,2} \hookrightarrow L^4$:[A.6]

$$\frac{\mathrm{d}}{\mathrm{d}\varepsilon}\big|_{\varepsilon=0} I(w+\varepsilon\,\eta)$$
$$= \frac{\mathrm{d}}{\mathrm{d}\varepsilon}\big|_{\varepsilon=0} I(u+\varepsilon\,\varphi, v+\varepsilon\,\psi)$$
$$= \frac{\theta}{2}\frac{\mathrm{d}}{\mathrm{d}\varepsilon}\big|_{\varepsilon=0}\langle \nabla_s u+\varepsilon\,\nabla_s\varphi+\tfrac{1}{2}(\nabla v+\varepsilon\,\nabla\psi)\otimes(\nabla v+\varepsilon\,\nabla\psi)\rangle$$
$$+ \frac{1}{24}\frac{\mathrm{d}}{\mathrm{d}\varepsilon}\big|_{\varepsilon=0}\langle \nabla^2 v-I+\varepsilon\,\nabla^2\psi\rangle$$
$$= \frac{\theta}{2}\frac{\mathrm{d}}{\mathrm{d}\varepsilon}\big|_{\varepsilon=0}\langle \underbrace{A+\varepsilon\,\nabla_s\varphi+\frac{\varepsilon}{2}\nabla v\otimes\nabla\psi+\frac{\varepsilon}{2}\nabla\psi\otimes\nabla v}_{\varepsilon B}$$
$$+\underbrace{\frac{\varepsilon^2}{2}\nabla\psi\otimes\nabla\psi}_{\varepsilon^2 C}\,\rangle+\frac{1}{12}\langle \nabla^2 v-I, \nabla^2\psi\rangle\,\mathrm{d}x$$
$$= \theta\langle A,B\rangle+\frac{1}{12}\int_\omega\langle \nabla^2 v-I, \nabla^2\psi\rangle$$
$$= \theta\langle A, \nabla_s\varphi+(\nabla v\otimes\nabla\psi)_{\mathrm{sym}}\rangle+\frac{1}{12}\langle \nabla^2 v-I, \nabla^2\psi\rangle.$$

We have then:

$$\delta I(w;\eta) = \theta\Big\langle \nabla_s u+\frac{1}{2}\nabla v\otimes\nabla v, \nabla_s\varphi+(\nabla v\otimes\nabla\psi)_{\mathrm{sym}}\Big\rangle$$
$$+\frac{1}{12}\langle \nabla^2 v-I, \nabla^2\psi\rangle. \qquad (A.16)$$

First Fréchet derivative: With estimates like those in Chapter 3 it is easy to see that the map (A.16) is linear in $(\varphi, \psi) \in X$ and bounded, so I is Gâteaux

---

A.6. Notice that because Q_2 vanishes on antisymmetric matrices, we could drop the symmetrisations but we leave them because they remind of the fact that the bilinear form is not elliptic on the whole space.

differentiable with differential given by the previous equation. It is also continuous in $w = (u, v)$. Hence I is Fréchet differentiable in X and by Lemma A.32 its differential $D_w I(w)$ is given by the integral in (A.16).

Second Fréchet derivative: In order to compute $D_w^2 I(w)$ we start with the iterated directional derivative (but we could also compute $\frac{\mathrm{d}}{\mathrm{d}\varepsilon}|_{\varepsilon=0} D I(u + \varepsilon\,\varphi_2, v + \varepsilon\,\psi_2; 0)[\varphi_1, \psi_1])$

$$\frac{\mathrm{d}^2}{\mathrm{d}\varepsilon_1\,\mathrm{d}\varepsilon_2}|_{\varepsilon_i=0} I(u + \varepsilon_1\,\varphi_1 + \varepsilon_2\,\varphi_2, v + \varepsilon_1\,\psi_1 + \varepsilon_2\,\psi_2) = \delta\,(\delta\,I(w; \varphi); \psi).$$

Again, the idea is that under suitable conditions the obtained map will be bilinear, symmetric and continuous and coincide with the second Fréchet differential. Though mostly trivial, the computations are tedious so we split them and make use of the fact that terms which are independent of ε_1 or ε_2 or quadratic in either one vanish when we differentiate. We begin with the second integrand:

$$\frac{\mathrm{d}^2}{\mathrm{d}\varepsilon_1\,\mathrm{d}\varepsilon_2}|_{\varepsilon_i=0} Q_2(\nabla^2 v + \varepsilon_1\,\nabla^2\psi_1 + \varepsilon_2\,\nabla^2\psi_2 - I)$$

$$= \frac{\mathrm{d}^2}{\mathrm{d}\varepsilon_1\,\mathrm{d}\varepsilon_2}|_{\varepsilon_i=0} \{Q_2(\nabla^2 v - I) + Q_2(\varepsilon_1\,\nabla^2\psi_1) + Q_2(\varepsilon_2\,\nabla^2\psi_2) +$$

$$+ 2\,Q_2[\nabla^2 v - I, \varepsilon_1\,\nabla^2\psi_1] + 2\,Q_2[\nabla^2 v - I, \varepsilon_2\,\nabla^2\psi_2]$$

$$+ 2\,Q_2[\varepsilon_1\,\nabla^2\psi_1, \varepsilon_2\,\nabla^2\psi_2]\}$$

$$= 2\,Q_2[\nabla^2\psi_1, \nabla^2\psi_2].$$

For the first integrand we find (again, terms independent of, or quadratic in, ε_1 or ε_2 vanish):

$$\frac{\mathrm{d}^2}{\mathrm{d}\varepsilon_1\,\mathrm{d}\varepsilon_2}|_{\varepsilon_i=0} Q_2\Big(\nabla_s u + \varepsilon_1\,\nabla_s\varphi_1 + \varepsilon_2\,\nabla_s\varphi_2$$

$$+ \frac{1}{2}\nabla(v + \varepsilon_1\,\psi_1 + \varepsilon_2\,\psi_2) \otimes \nabla(v + \varepsilon_1\,\psi_1 + \varepsilon_2\,\psi_2)\Big)$$

$$= \frac{\mathrm{d}^2}{\mathrm{d}\varepsilon_1\,\mathrm{d}\varepsilon_2}|_{\varepsilon_i=0} Q_2(F + \varepsilon_1 A_1 + \varepsilon_2 A_2 + \varepsilon_1^2(\ldots) + \varepsilon_2^2(\ldots)$$

$$+ \varepsilon_1\varepsilon_2\,(\nabla\psi_1 \otimes \nabla\psi_2)_{\mathrm{sym}})$$

$$= 2\,Q_2[F, (\nabla\psi_1 \otimes \nabla\psi_2)_{\mathrm{sym}}] + 2\,Q_2[A_1, A_2],$$

where $F = \nabla_s u + \frac{1}{2} \nabla v \otimes \nabla v$, $A_1 = \nabla_s \varphi_1 + (\nabla v \otimes \nabla \psi_1)_{\text{sym}}$ and $A_2 = \nabla_s \varphi_2 + (\nabla v \otimes \nabla \psi_2)_{\text{sym}}$. We bring both computations together and obtain the map:

$$[(\varphi_1, \psi_1), (\varphi_2, \psi_2)] \mapsto \theta \int_\omega Q_2 \left[\nabla_s u + \frac{1}{2} \nabla v \otimes \nabla v, (\nabla \psi_1 \otimes \nabla \psi_2)_{\text{sym}} \right]$$

$$+ \theta \int_\omega Q_2 [\nabla_s \varphi_1 + (\nabla v \otimes \nabla \psi_1)_{\text{sym}},$$

$$\nabla_s \varphi_2 + (\nabla v \otimes \nabla \psi_2)_{\text{sym}}]$$

$$+ \frac{1}{12} \int_\omega Q_2 [\nabla^2 \psi_1, \nabla^2 \psi_2].$$

This is clearly linear on (φ_i, ψ_i) and because it is also continuous on (u, v), it is in fact the second Fréchet differential of I by Lemma A.33. Finally, evaluation at any $w \in X$ and $\theta = 0$ yields

$$D_w^2 I(w; 0)[(\varphi_1, \psi_1), (\varphi_2, \psi_2)] = \frac{1}{12} \int_\omega Q_2 [\nabla^2 \psi_1, \nabla^2 \psi_2]. \qquad (A.17)$$

Partial derivatives: Finally, at the risk of being a bit repetitious, we collect the partial derivatives of I, for their use in Theorem 3.10. Setting $\psi = 0$ in (A.16), we have that the first variation of I along $\varphi \in X_u$ (Definition 2.12) is

$$\frac{\mathrm{d}}{\mathrm{d}\varepsilon}\Big|_{\varepsilon=0} I(u + \varepsilon \varphi, v) = \theta \langle \nabla_s u + \frac{1}{2} \nabla v \otimes \nabla v, \nabla_s \varphi \rangle.$$

This map is linear in φ and bounded and it is also continuous on $u \in X_u$. Hence I is Fréchet differentiable wrt. $u \in X_u$ and by Lemma A.32 its differential $\partial_u I$ at (u, v) is given by the previous integral. Analogously we set $\varphi = 0$ in (A.16) and obtain for $\psi \in X_v$:

$$\frac{\mathrm{d}}{\mathrm{d}\varepsilon}\Big|_{\varepsilon=0} I(u, v + \varepsilon \psi) = \theta \langle \nabla_s u + \frac{1}{2} \nabla v \otimes \nabla v, (\nabla v \otimes \nabla \psi)_{\text{sym}} \rangle$$

$$+ \frac{1}{12} \langle \nabla^2 v - I, \nabla^2 \psi \rangle$$

And in the same manner as above we conclude that this expression yields the derivative $\partial_v I(u, v; \theta)$.

Appendix B

Notation

$\simeq, \lesssim$ denote equality, resp. inequality, up to a constant which doesn't depend on the quantities involved.

$x = (x', x_3) \in \mathbb{R}^3, x' \in \mathbb{R}^2.$

$[A] := \sqrt{A^\top A}$ for any matrix A.

$[N] := \{n \in \mathbb{N} : n < N\}.$

$A_{\text{sym}} = \text{sym}\, A = \frac{1}{2}(A + A^\top)$, the symmetric part of matrix A.

$A_{\text{ant}} = \text{ant}\, A = \frac{1}{2}(A - A^\top)$, the antisymmetric part of matrix A.

$O(n)$ Orthogonal group: Real $n \times n$ matrices A such that $A^\top A = I$.

$SO(n)$ Special orthogonal group. Real $n \times n$ matrices A such that $A^\top A = I$ and $\det A = 1$.

$so(n)$ The set of real $n \times n$ antisymmetric matrices.

$\nabla f = (\partial_1 f, \partial_2 f, \partial_3 f)^\top$, a column vector, for $f : \mathbb{R}^3 \to \mathbb{R}$.

$(\nabla y)_{ij} = y_{i,j} = \partial_j y_i, i, j \in \{1, 2, 3\}$, each row $\nabla^\top y_i$, if $y : \mathbb{R}^3 \to \mathbb{R}^3$.

$(\nabla' y)_{\rho\tau} = y_{\rho,\tau}, \rho, \tau \in \{1, 2\}$ for any $y : \mathbb{R}^3 \to \mathbb{R}^3$ is the 2×2 gradient or Jacobi matrix of y wrt. the first two variables.

$\nabla_s u = \frac{1}{2}(\nabla u + \nabla^\top u)$, the symmetrised gradient of $u : \mathbb{R}^2 \to \mathbb{R}^2$.

$\nabla_h y = \left(\partial_1 y, \partial_2 y, \frac{1}{h}\partial_3 y\right)$ with $y : \Omega \subset \mathbb{R}^3 \to \mathbb{R}^3$ a deformation.

$\nabla^2 v$, the Hessian matrix of $v: \mathbb{R}^n \to \mathbb{R}$.

Δ^2, the bilaplacian operator. In $\mathbb{R}^2$: $\Delta^2 = \partial_1^4 + 2\,\partial_1^2 \partial_2^2 + \partial_2^4$.

$\hat{G} := G_{\alpha\beta}\, e_\alpha \otimes e_\beta \in \mathbb{R}^{3\times3}$, $G \in \mathbb{R}^{2\times2}$, $\alpha, \beta \in \{1, 2\}$ and $e_1, e_2 \in \mathbb{R}^3$ are the standard basis vectors.

$\hat{\nabla} u := (\nabla' u)_{\alpha\beta}\, e_\alpha \otimes e_\beta \in \mathbb{R}^{3\times3}$ for $u: \omega \subset \mathbb{R}^2 \to \mathbb{R}^2$, and $\alpha, \beta \in \{1, 2\}$, e_1, $e_2 \in \mathbb{R}^3$ the standard basis vectors.

$\hat{\nabla} b := (\nabla' b)_{\alpha\beta}\, e_\alpha \otimes e_\beta \in \mathbb{R}^{3\times3}$ for $b: \omega \subset \mathbb{R}^2 \to \mathbb{R}^3$, and $\alpha \in \{1, 2, 3\}$, $\beta \in \{1, 2\}$, $e_1, e_2, e_3 \in \mathbb{R}^3$ the standard basis vectors.

$\hat{\nabla} v := (\partial_1 v, \partial_2 v, 0)^\top \in \mathbb{R}^3$ for $v: \omega \subset \mathbb{R}^2 \to \mathbb{R}$.

$\check{B} \in \mathbb{R}^{2\times2}$ is the matrix resulting from the deletion of the third row and column of any $B \in \mathbb{R}^{3\times3}$

A quadratic form $Q(\cdot)$ has the associated unique bilinear form $Q[\cdot, \cdot]$.

$\langle F, G \rangle := \int_\omega Q_2[F, G]$ and $\langle F \rangle := \langle F, F \rangle = \int_\omega Q_2(F)$ for all $F, G \in L^2(\omega;$ $\mathbb{R}^{2\times2})$. Sometimes we also write $\|F\|_{Q_2}^2 = \langle F \rangle$.

$\|v\|_{k,p,\Omega} = \|v\|_{W^{k,p}(\Omega)}$. We will omit the domain when it is clear from the context.

$A_\theta := \nabla_s u_\theta + \frac{1}{2}\nabla v_\theta \otimes \nabla v_\theta$, mostly in Section 2.4.

$(f)_\omega := \frac{1}{|\omega|} \int_\omega f(x')\,\mathrm{d}x'$ is the average of f over ω.

μ, λ are the Lamé constants of an isotropic hyperelastic material.

w.s.l.s.c. "weakly sequentially lower semicontinuous".

wrt. "with respect to".

dof "degree of freedom".

$(\dagger),(\ddagger),(\star)$ Are references to equations valid only in the local scope (the innermost environment, section, etc. in which they appear).

References

ABH+15 Martin S. Alnaes, Jan Blechta, Johan Hake, August Johansson, Benjamin Kehlet, Anders Logg, Chris Richardson, Johannes Ring, Marie E. Rognes, and Garth N. Wells. The FEniCS Project Version 1.5. *Archive of Numerical Software*, 3(100), 2015.

ABP88 Emilio Acerbi, Giuseppe Buttazzo, and Danilo Percivale. Thin inclusions in linear elasticity: a variational approach. *Journal für die reine und angewandte Mathematik*, 386:99–115, 1988.

ABP91 Emilio Acerbi, Giuseppe Buttazzo, and Danilo Percivale. A variational definition of the strain energy for an elastic string. *Journal of Elasticity*, 25(2):137–148, mar 1991.

ABP94 G. Anzellotti, S. Baldo, and Danilo Percivale. Dimension reduction in variational problems, asymptotic development in Γ-convergence and thin structures in elasticity. *Asymptotic Analysis*, 9(1):61–100, jan 1994.

ADD12 Virginia Agostiniani, Gianni Dal Maso, and Antonio Desimone. Linear elasticity obtained from finite elasticity by Γ-convergence under weak coerciveness conditions. *Annales de l'Institut Henri Poincaré. Analyse non linéaire*, 29:715–735, 2012.

AF03 Robert A. Adams and John J. F. Fournier. *Sobolev Spaces*. Number 140 in Pure and Applied Mathematics. Academic Press, Department of Mathematics, The University of British Columbia Vancouver, Canada, 2nd edition, 2003.

Alt12 Hans Wilhelm Alt. *Lineare Funktionalanalysis*. Springer, 6th edition, 2012.

AMZ02 Douglas N. Arnold, Alexandre L. Madureira, and Sheng Zhang. On the range of applicability of the Reissner-Mindlin and Kirchhoff-Love plate bending models. *Journal of elasticity and the physical science of solids*, 67(3):171–185, jun 2002.

Bar13 S. Bartels. Approximation of Large Bending Isometries with Discrete Kirchhoff Triangles. *SIAM Journal on Numerical Analysis*, 51(1):516–525, jan 2013.

Bar15 Sören Bartels. *Numerical Methods for Nonlinear Partial Differential Equations*, volume 47 of *Springer Series in Computational Mathematics*. Springer International Publishing, Cham, 2015.

Bar16 Sören Bartels. Numerical solution of a Föppl-von Kármán model. 2016.

BCDM02 Hafedh Ben Belgacem, Sergio Conti, Antonio DeSimone, and Stefan Müller. Energy scaling of compressed elastic films - three-dimensional elasticity and reduced theories. *Archive for Rational Mechanics and Analysis*, 164(1):1–37, aug 2002.

BD98 Andrea Braides and Anneliese Defranceschi. *Homogenization of Multiple Integrals*. Clarendon Press, 1998.

Bha09 Rajendra Bhatia. *Positive Definite Matrices*. Princeton Series in Applied Mathematics. Princeton University Press, jan 2009.

BNRS17 Susanne C. Brenner, Michael Neilan, Armin Reiser, and Li-Yeng Sung. A C^0 interior penalty method for a von Kármán plate. *Numerische Mathematik*, 135(3):803–832, mar 2017.

BP90 I. Babuška and J. Pitkäranta. The plate paradox for hard and soft simple support. *SIAM Journal on Mathematical Analysis*, 21(3):551–576, may 1990.

Bra06 Andrea Braides. A handbook of Γ-convergence. In M. Chipot and P. Quittner, editors, *Stationary Partial Differential Equations*, volume 3 of *Handbook of Differential Equations*, pages 101–213. Elsevier, 2006.

BS92 Ivo Babuška and Manil Suri. On Locking and Robustness in the Finite Element Method. *SIAM Journal on Numerical Analysis*, 29(5):1261–1293, 1992.

BS08 Susanne C. Brenner and L. Ridgway Scott. *The Mathematical Theory of Finite Element Methods*. Number 15 in Texts in Applied Mathematics. Springer New York, New York, NY, 3rd edition, 2008.

CFM05 Sergio Conti, Daniel Faraco, and Francesco Maggi. A new approach to counterexamples to L^1 estimates: Korn's inequality, geometric rigidity, and regularity for gradients of separately convex functions. *Archive for Rational Mechanics and Analysis*, 175(2):287–300, feb 2005.

Cia88 Philippe G. Ciarlet. *Mathematical Elasticity: Three-Dimensional Elasticity*, volume 1 of *Studies in Mathematics and Its Applications*. Elsevier, 1988.

Cia97 Philippe G. Ciarlet. *Mathematical Elasticity: Theory of Plates*, volume 2 of *Studies in Mathematics and Its Applications*. North-Holland, 1997.

Cia05 Philippe G. Ciarlet. *An Introduction to Differential Geometry with Applications to Elasticity*. Springer, City University of Hong Kong, aug 2005.

CM08 Sergio Conti and Francesco Maggi. Confining thin elastic sheets and folding paper. *Archive for Rational Mechanics and Analysis*, 187(1):1–48, jan 2008.

Con04 Sergio Conti. *Low-Energy Deformations of Thin Elastic Plates: Isometric Embeddings and Branching Patterns*. Habilitationsschreiben, Leipzig University, 2004.

CS06 Sergio Conti and Ben Schweizer. Rigidity and Γ-convergence for solid-solid phase transitions with SO(2) invariance. *Communications on Pure and Applied Mathematics*, 59(6):830–868, jun 2006.

Dac07 Bernard Dacorogna. *Direct Methods in the Calculus of Variations*. Number 78 in Applied Mathematical Sciences. Springer, 2nd edition, jan 2007.

dB17a Miguel de Benito Delgado. Implementation of a generalised von Kármán model for multilayered plates. `https://bitbucket.org/mdbenito/lvk`, nov 2017.

dB17b Miguel de Benito Delgado. Implementation of a nonlinear Kirchhoff plate model. `https://bitbucket.org/mdbenito/nonlinear-kirchhoff`, aug 2017.

dB17c Miguel de Benito Delgado. Implementation of Hermite elements for FEniCS. `https://bitbucket.org/mdbenito/hermite`, feb 2017.

DFMŠ17 Bernard Dacorogna, Nicola Fusco, Stefan Müller, and Vladimír Šverák. *Vector-Valued Partial Differential Equations and Applications*. Number 2179 in Lecture Notes in Mathematics. Springer, Cetraro, Italy 2013, 2017.

DNP02 G. Dal Maso, M. Negri, and Danilo Percivale. Linearized Elasticity as Γ-Limit of Finite Elasticity. *Set-Valued Analysis*, 10(2-3):165–183, 2002.

FJM02 Gero Friesecke, Richard D. James, and Stefan Müller. A theorem on geometric rigidity and the derivation of nonlinear plate theory from three-dimensional elasticity. *Communications on Pure and Applied Mathematics*, 55(11):1461–1506, 2002.

FJM06 Gero Friesecke, Richard D. James, and Stefan Müller. A hierarchy of plate models derived from nonlinear elasticity by Γ-convergence. *Archive for Rational Mechanics and Analysis*, 180(2):183–236, may 2006.

FL07 Irene Fonseca and Giovanni Leoni. *Modern Methods in the Calculus of Variations: L^p Spaces*. Springer Monographs in Mathematics. Springer, 2007.

FP15 Lior Falach, Roberto Paroni, and Paolo Podio-Guidugli. A justification of the Timoshenko beam model through Γ-convergence. *Analysis and Applications*, 15(02):261–277, aug 2015.

Gal01 Jean Gallier. *Geometric Methods and Applications for Computer Science and Engineering*. Number 38 in Texts in Applied Mathematics. Springer-Verlag, New York, 2001.

GKC+17 Klaus Greff, Aaron Klein, Martin Chovanec, Frank Hutter, and Jürgen Schmidhuber. The Sacred Infrastructure for Computational Research. *Proceedings of the 16th Python in Science Conference*, pages 49–56, 2017.

GRS07 Christian Grossmann, Hans-Görg Roos, and Martin Stynes. *Numerical Treatment of Partial Differential Equations*. Universitext. Springer Berlin Heidelberg, Berlin, Heidelberg, 2007.

HJ12 Roger A. Horn and Charles R. Johnson. *Matrix Analysis*. Cambridge University Press, 2nd edition, 2012.

HKO08 Peter Howell, Gregory Kozyreff, and John Ockendon. *Applied Solid Mechanics*. Number 43 in Cambridge Texts in Applied Mathematics. Cambridge University Press, dec 2008.

Hor11 Peter Hornung. Approximation of flat $W^{2,2}$ isometric immersions by smooth ones. *Archive for Rational Mechanics and Analysis*, 199(3):1015–1067, mar 2011.

JS13 Martin Jesenko and Bernd Schmidt. Closure and commutability results for Gamma-Limits and the geometric linearization and homogenization of multi-well energy functionals. *ArXiv:1308.0963 [cond-mat]*, aug 2013.

KP+16 Thomas Kluyver, Benjamin Ragan-Kelley, Fernando Pérez, Brian Granger, Matthias Bussonnier, Jonathan Frederic, Kyle Kelley, Jessica Hamrick, Jason Grout, Sylvain Corlay, Paul Ivanov, Damián Avila, Safia Abdalla, and Carol Willing. Jupyter Notebooks – a publishing format for reproducible computational workflows. In F. Loizides and B. Schmidt, editors, *Positioning and Power in Academic Publishing: Players, Agents and Agendas*, pages 87–90. IOS Press, 2016.

Lan69 Serge Lang. *Analysis II*. Addison-Wesley Pub. Co., jun 1969.

LMP11 Marta Lewicka, L. Mahadevan, and Mohammad Reza Pakzad. The Föppl-von Kármán equations for plates with incompatible strains. *Proceedings of the Royal Society of London. Series A. Mathematical, Physical and Engineering Sciences*, 467(2126):402–426, 2011.

LMP14 Marta Lewicka, L. Mahadevan, and Mohammad Reza Pakzad. Models for elastic shells with incompatible strains. *Proceedings of the Royal Society A: Mathematical, Physical and Engineering Science*, 470(2165):20130604, aug 2014.

LP09 Marta Lewicka and Reza Pakzad. The infinite hierarchy of elastic shell models: some recent results and a conjecture. *ArXiv:0907.1585 [math]*, jul 2009.

LR95 Hervé Le Dret and Annie Raoult. The nonlinear membrane model as variational limit of nonlinear three-dimensional elasticity. *Journal de mathématiques pures et appliquées*, 74(6):549–578, 1995.

MH94 Jerrold E. Marsden and Thomas J. R. Hughes. *Mathematical Foundations of Elasticity*. Dover civil and mechanical engineering, California Institute of Technology, Pasadena, Reprint of the 1983 edition, 1994.

MN16a Gouranga Mallik and Neela Nataraj. A nonconforming finite element approximation for the von Kármán equations. *ESAIM: Mathematical Modelling and Numerical Analysis*, 50(2):433–454, mar 2016.

MN16b Gouranga Mallik and Neela Nataraj. Conforming finite element methods for the von Kármán equations. *Advances in Computational Mathematics*, 42(5):1031–1054, oct 2016.

MP05 Stefan Müller and Mohammad Reza Pakzad. Regularity properties of isometric immersions. *Mathematische Zeitschrift*, 251(2):313–331, jul 2005.

Orn62 Donald Ornstein. A non-inequality for differential operators in the L^1 norm. *Archive for Rational Mechanics and Analysis*, 11(1):40–49, jan 1962.

Ort04 Christoph Ortner. Γ-Limits of Galerkin Discretizations with Quadrature. Technical Report 04/26, Oxford University Computing Laboratory, Numerical Analysis Group, dec 2004.

P15 Roberto Paroni and Paolo Podio-Guidugli. On variational dimension reduction in structure mechanics. *Journal of Elasticity*, 118(1):1–13, jan 2015.

Pak04 Mohammad Reza Pakzad. On the Sobolev space of isometric immersions. *Journal of Differential Geometry*, 66(1):47–69, jan 2004.

Pra01 Gangan Prathap. Finite element analysis as computation. 2001.

PT07 Roberto Paroni, Paolo Podio-Guidugli, and Giuseppe Tomassetti. A justification of the Reissner-Mindlin plate theory through variational convergence. *Analysis and Applications*, 05(02):165–182, apr 2007.

PT17 Roberto Paroni and Giuseppe Tomassetti. Linear models for thin plates of polymer gels. *Mathematics and Mechanics of Solids*, mar 2017.

PW60 L. E. Payne and H. F. Weinberger. An optimal Poincaré inequality for convex domains. *Archive for Rational Mechanics and Analysis*, 5(1):286–292, jan 1960.

Qua12 Alessio Quaglino. *Membrane Locking in Discrete Shell Theories*. Doctoral dissertation, Georg-August-Universität Göttingen, Göttingen, may 2012.

Sch07a Bernd Schmidt. Minimal energy configurations of strained multi-layers. *Calculus of Variations and Partial Differential Equations*, 30(4):477–497, dec 2007.

Sch07b Bernd Schmidt. Plate theory for stressed heterogeneous multilayers of finite bending energy. *Journal de Mathématiques Pures et Appliquées*, 88(1):107–122, jul 2007.

Sch09 Bernd Schmidt. On the derivation of linear elasticity from atomistic models. *Networks and heterogeneous media*, 4(4):789–812, dec 2009.

Sub18 Vivek Ratnavel Subramanian. Omniboard: a web-based dashboard for Sacred. sep 2018.

TM05 Roger Temam and Alain Miranville. *Mathematical Modeling in Continuum Mechanics*. Cambridge University Press, 2nd edition, 2005.

Ves12 Matthias Vestner. *Effective Theories for Internally Stressed Bodies Derived from Nonlinear Elasticity by Γ-Convergence*. Diplomarbeit, Technische Universität München, Munich, dec 2012.

Wer07 Dirk Werner. *Funktionalanalysis*. Springer-Lehrbuch. Springer-Verlag, Berlin Heidelberg, 6th edition, 2007.

In der Reihe *Augsburger Schriften zur Mathematik, Physik und Informatik,* herausgegeben von Prof. Dr. B. Aulbach, Prof. Dr. F. Pukelsheim, Prof. Dr. W. Reif, Prof. Dr. B. Schmidt, Prof. Dr. D. Vollhardt,

sind bisher erschienen:

1	Martin Mißlbeck	Entwicklung eines schnellen Spektralradiometers und Weiterentwicklung herkömmlicher Messverfahren zur Messung der solaren UV-Strahlung
		ISBN 978-3-8325-0208-9, 2003, 139 S. 40.50 €
2	Bernd Reinhard	Dynamisches Trapping in modulierten monotonen Potentialen
		ISBN 978-3-8325-0516-5, 2004, 154 S. 40.50 €
3	Cosima Schuster	Physikerinnen stellen sich vor - Dokumentation der Deutschen Physikerinnentagung 2003
		ISBN 978-3-8325-0520-2, 2004, 164 S. 40.50 €
4	Udo Schwingenschlögl	The Interplay of Structural and Electronic Properties in Transition Metal Oxides
		ISBN 978-3-8325-0530-1, 2004, 174 S. 40.50 €
5	Marianne Leitner	Zero Field Hall-Effekt für Teilchen mit Spin $1/2$
		ISBN 978-3-8325-0578-3, 2004, 81 S. 40.50 €
6	Georg Keller	Realistic Modeling of Strongly Correlated Electron Systems
		ISBN 978-3-8325-0970-5, 2005, 150 S. 40.50 €
7	Niko Tzoukmanis	Local Minimizers of Singularly Perturbed Functionals with Nonlocal Term
		ISBN 978-3-8325-0650-6, 2004, 154 S. 40.50 €

35	Johanna Kerler-Back	Dynamic iteration and model order reduction for magneto-quasistatic systems
		ISBN 978-3-8325-4910-7, 2019, 175 S. 35.50 €
36	Veronika Antonie Auer-Volkmann	Eigendamage: An Eigendeformation Model for the Variational Approximation of Cohesive Fracture
		ISBN 978-3-8325-4969-5, 2019, 151 S. 39.00 €
37	Miguel de Benito Delgado	Effective two dimensional theories for multi-layered plates
		ISBN 978-3-8325-4984-8, 2019, 153 S. 38.00 €

Alle erschienenen Bücher können unter der angegebenen ISBN im Buchhandel oder direkt beim Logos Verlag Berlin (www.logos-verlag.de, Fax: 030 - 42 85 10 92) bestellt werden.